Ask a Philosopher

Ask a Philosopher

Answers to
Your Most Important
and Most Unexpected
Questions

Ian Olasov

THOMAS DUNNE BOOKS
NEW YORK

First published in the United States by Thomas Dunne Books, an imprint of
St. Martin's Publishing Group

ASK A PHILOSOPHER. Copyright © 2020 by Ian Olasov. All rights reserved. Printed in the
United States of America. For information, address St. Martin's Publishing Group,
120 Broadway, New York, NY 10271.

www.thomasdunnebooks.com

Design by Jonathan Bennett

Library of Congress Cataloging-in-Publication Data

Names: Olasov, Ian, author.
Title: Ask a philosopher : answers to your most important and most
    unexpected questions / Ian Olasov.
Identifiers: LCCN 2020019308 | ISBN 9781250756176 (hardcover) |
    ISBN 9781250756183 (ebook)
Subjects: LCSH: Philosophy—Miscellanea.
Classification: LCC B68 .O427 2020 | DDC 100—dc23
LC record available at https://lccn.loc.gov/2020019308

ISBN 978-1-250-75617-6 (hardcover)
ISBN 978-1-250-75618-3 (ebook)

Our books may be purchased in bulk for promotional, educational, or business use.
Please contact your local bookseller or the Macmillan Corporate and Premium Sales
Department at 1-800-221-7945, extension 5442, or by email at
MacmillanSpecialMarkets@macmillan.com.

First Edition: 2020

10 9 8 7 6 5 4 3 2 1

For the questioners

Contents

Preface

One Saturday morning in April 2016—a little cold, a little wet, surrounded by flowers—I unfolded a table at the Grand Army Plaza farmers market, across the street from the central branch of the Brooklyn Public Library. For the next few hours, a handful of philosophy professors and grad students and I sat behind a banner that said ASK A PHILOSOPHER, and we waited for people to talk to us. Before long, they did—about God, the presidential election, Ayn Rand, keeping fish as pets, moral education, free will, destiny, the meaning of life, and a few other things. So we set up the booth again, and then another time, and then a few more times after that. In the years since, we've traveled all over New York City, to farmers markets, stores, subway stations, parks, book festivals, and street fairs. It's hard to describe how rewarding the whole experience has been. I've met, for a few seconds or a couple of hours, thousands of weird, friendly, cranky, curious, lonely, unhinged, effervescent, wise people of every conceivable demographic category. Each new installment brings new questions, new insights, new stories.

I started the Ask a Philosopher booth because I want philosophy to be responsive to the needs of ordinary people. It's important to enable and encourage people to find out about the problems that preoccupy professional philosophers, but it's at least as important to enable and encourage philosophers to find out about—and help with—the problems that preoccupy everyone else.

This book offers an answer—or a fragment of an answer—to a

bunch of the stickiest questions posed by visitors to the booth.* The questions reflect the enormous range of what we care about. Sidebars throughout paint the little scenes that help make each booth and each question memorable. You'll occasionally see lines (✎), which represent the responses of an imaginary interlocutor. Feel free to read the book in whatever order you want.

Some of the magic of the booth is hard to capture on the page— the spontaneity, the interactivity, the funky theater of springing philosophy on people who have other things on their minds. But the book captures something. I hope the discussion below gives the sense that there's a way of making philosophy work for each of us, that philosophy can be both perfectly continuous with everyday life and also utterly transporting.

✝

I believe each claim I make in this book. I also believe that the book contains some false claims. The first I believe because I've written this stuff sincerely, the second because philosophy is hard, and I have something approximating a healthy appreciation for my own limits.

You might see where this is going. These beliefs are inconsistent; they can't all be true. And usually, if I discover that some of my beliefs are inconsistent, I'll revise the beliefs until they no longer are. After all, beliefs are things that you reason with, and reasoning with inconsistent information is a headache. And if some set of beliefs can't all be right, at least one of them has to be wrong. But even if I went back and checked my work, that wouldn't help. I would still be making a bunch of claims, and I would still want to acknowledge that I'm sure I blew it *somewhere*. We have a paradox on our hands.

Luckily—maybe a little too luckily—this is a paradox with a nice, tidy solution. The solution is that there's *belief* and there's belief. Or rather, there's full belief, and there's belief by degrees. If I fully believe some inconsistent claims, I have work to do. But if I believe a bunch

---

* More or less. Some of the questions were posed directly by the visitors, some of them arose in the course of conversations with visitors.

[XII]

of things merely to a high degree, I can also believe to a high degree that one of them is wrong.[*]

This all illustrates a few important things.

First, a lot of philosophy arises from poking at the inconsistencies in our own beliefs. Sometimes we'll be able to reason our way out of them, but sometimes we'll have to find a way to balance our beliefs between them.

Second, I believe what I've written in this book, but not 100 percent. When it feels like I'm verging on insincerity or overconfidence, it helps to remind myself that I'm at least as interested in stimulating fruitful philosophical inquiry as I am in sharing with you the correct answers to some philosophical questions.

Third, like this discussion of the paradox of the preface, the discussions of the problems in this book are much shorter than they could be. At every turn, there are reasonable objections I don't consider, alternative hypotheses I don't explore, details I gloss over. But you're likely better than I am at coming up with objections and alternatives to my own ideas, so why should I try to do it for you? In any case, I want reading this book to be mostly fun and entirely nonboring.

Fourth, like all of the above, much of this book is borrowed. A lot of what I say here has already been said by someone else. To keep things breezy enough, I've limited all the attributions and suggestions for further reading to an appendix.

Lastly, all of this breeziness might give the impression that the whole thing seems suspiciously easy. Let me be clear that philosophy is hard—full of doubt and dead ends, never on solid ground, always at the edge of nonsense or irrelevance. But philosophy is *just* hard. It's not impossible.

---

[*] To see this, imagine that your beliefs are like probabilities. So if you 95 percent believe one hundred independent things, you'll also 99 percent believe that one of them is wrong.

PART I

Cosmic Questions

What Is Philosophy?

On the first day of class, when I'm trying to give my students a sense of what philosophy is all about, I give them a bunch of examples of philosophical questions. Inevitably, someone says something like, "Oh, you mean questions that you can't really answer." But I would resist this characterization. For starters, a lot of the questions that *used* to be considered philosophical (the question whether matter is infinitely divisible, for instance) have become scientific questions. Who's to say that the questions in this book won't become scientific questions eventually? (I think some of these questions are *already* settled science, but scientists themselves are reluctant to speak out about them for whatever reason.) But there's something to this. There's nothing like a consensus—among professional philosophers or the world at large—about the correct answers to philosophical questions. And this lack of consensus isn't (just) due to the fact that some people haven't thought things through carefully enough; in some cases, maximally reasonable, well-informed people can disagree about their answers.

Another suggestive but not quite right idea about what makes a question philosophical is illustrated by a funny thing that happens at the Ask a Philosopher booth. Usually, at some point over the course of the day, someone will see the sign and ask us a question about astrology or dream interpretation or astral projection or who shot JFK. It takes a bit of work to bring these discussions back to questions that I regard as philosophical. (What does the popularity of astrology tell us about the role of storytelling in our lives? What would make a dream interpretation correct? When is it reasonable to believe a conspiracy theory?) But why do people think that these questions are philosophical in the first place? In part, it's just because philosophers as a whole haven't put too much thought into how they communicate their work to the public. But I think it's also because people have the sense, correctly, that philosophy is where you go to hear out ideas that aren't taken seriously elsewhere. This is true in the sense that philosophical arguments often rely on artificial or outlandish thought experiments.

(<digression> Some of my favorites:

The Trolley Problem: If you saw a trolley headed toward five people tied to a track and you could flip a switch to divert it toward a single person, should you? If you saw a trolley headed toward five people, and you could push a large person in the path of the trolley to stop it, should you? Should the two questions receive the same answer, and if not, why not?

The Veil of Ignorance: Imagine that you temporarily knew everything you could ever want to know about the society you live in, except for who you are in that society. What changes would you make to the society's basic laws and institutions? Since you couldn't exploit any special bargaining power to advance your own interests, would these changes necessarily make for a more just society?

Twin Earth: Would "water" mean the same thing in a world that looked exactly like our own but where the stuff people called water was made of something other than H_2O?

The Invisible Gardener: Is there a difference between a garden tended by a gardener who is impossible

to detect and a garden tended by no gardener
at all?

The New Riddle of Induction: If English had a word "grue,"
which meant "first observed before 2030 and is green or
not observed before 2030 and is blue," would the fact
that all the fresh grass you have ever seen is grue give
you reason to believe that all grass is grue?

Gödel and Schmidt: If everything you believe about the
person you call Kurt Gödel was actually true of someone you've never heard of named Schmidt, do you have
a bunch of false beliefs about Gödel or a bunch of true
beliefs about Schmidt?

The Knowledge Argument: If someone had lived her
whole life in black and white, but knew everything
there was to know about the physics and psychology and neuroscience of color perception, what, if
anything, would she learn the first time she saw a red
apple?

Freeze World: In a universe divided into three parts, one
of which seems to outsiders to stand perfectly still for
five minutes once every year, one of which seems to
freeze for five minutes once every two years, and one
of which seems to freeze every three years, does five
minutes pass without anything changing once every six
years?

The Floating Man: If you were born without any of your
senses, would you still be aware of yourself?

The Ring of Gyges: If you had a ring that made you
invisible when you wear it, would it make you an awful
person? What would stop you from stealing, cheating,
stalking, and generally doing all the selfish things you
could get away with?

Dennett's "Where Am I?": If your brain remotely controlled the rest of your body through tiny radio transmitters placed on each of your nerve endings, would you be where your brain is or where your body is?

Gettier Cases: If, unbeknownst to you, someone has put your phone on silent and you hear your ringtone coming from another nearby phone, but at the same time, by complete coincidence, someone is actually calling you, do you know that you're getting a call?

Radical Translation: If you are with someone speaking a language that is, as far as you know, completely unrelated to any language you speak and they point to a rabbit and say "Gavagai," how do you know that "Gavagai" means *rabbit*, rather than *undetached rabbit part*, or *rabbit time-slice*, or *the property of being a rabbit*?
</digression>)

It's also true in the sense that some conclusions widely held among philosophers (that no one has conscious experiences, that the passage of time is an illusion, that no one knows anything) are ideas that we refuse to entertain in everyday life. In philosophy, at least when it's relevant, it's not enough just to dismiss these ideas out of hand; you have to reason about them.

Here's a way of thinking about philosophy that works pretty well: if there's no consensus about what methods or sources of evidence we should use to study some question, it's philosophical. This is true of all the philosophical questions discussed in this book I think. It would also explain why people have the sense that philosophical questions are unanswerable, why questions leave philosophy over time, why there are philosophical questions to ask about every subject, and why open-mindedness is such an important virtue in philosophy.

Still, this isn't quite right. There are pretty established methods for doing research in logic and the history of philosophy, which are part of philosophy if anything is, and there is no consensus about how to study some difficult problems in physics and history and psychology. But it's the best I've got. If you know a better way of explaining what philosophy is, send me an email.

When we set up the Ask a Philosopher booth, we put out a bowl full of philosophical questions, a bowl full of thought experiments, and a bowl of candy. Toward the end of a hot summer day at the booth, the candy dish ran dry. A visitor looked at the empty candy bowl and asked, "Is this some kind of metaphor for philosophy?" That one hurt.

Why Is There Anything
Instead of Nothing?

When I was a toddler, I had a memorable temper tantrum. I wanted fried eggs for breakfast, but I thought fried eggs were called scrambled eggs. So I asked for scrambled eggs, got them, and had a fit. When my parents offered me fried eggs, my fit continued. I didn't just want fried eggs; I also wanted fried eggs to be called scrambled eggs. I was, perhaps not for the last time, asking for more than it was possible for my parents to deliver.

So why is there anything at all? On its face, this appears to be a request for a causal explanation. So you could paraphrase it as: what caused the first things to exist? It's logically possible that the first things caused themselves to exist or that something that came after them caused them to exist. But let's set those possibilities aside. (One reason to do so is that they stretch the concept of causation, perhaps to breaking. Another is that if things could cause themselves to exist or be caused to exist by later events, it's unclear why this doesn't happen all the time.) In that case, the only direct answer we're left with is that something preceded the first things and caused them to exist. But that's absurd. If something preceded the first things, they wouldn't be the first things. It's like asking, "What's the name of Bill Clinton's third son?" The question has no answer, not because it's hard but because it assumes something false.

✎ Of course there's no causal explanation for the beginning of the world. But that's not what I'm interested in.

It's true that causal explanations aren't the only types of explanations. We can explain mathematical facts (like the fact that $2 + 2 = 4$) by deducing them in a humanly intelligible way from intuitive axioms (like the axioms of Peano arithmetic[*]); we can explain special laws of nature (like Kepler's laws of planetary motion) by showing how they follow from more general laws of nature (like Newton's law of gravity and laws of motion); we can explain an action or belief in terms of the reasons in favor of it; we can explain a trait of an organism in terms of its function; we can explain hard-to-understand ideas by rephrasing them in familiar terms or by using vivid analogies or examples. But the question evidently isn't asking for the fact that stuff exists to be deduced from mathematical principles or laws of nature or rephrased in a way that makes sense—or anything like that. So it's asking for an explanation, but not any kind of explanation anyone has heard of or could recognize. Like asking for a fried egg that's called a scrambled egg, this seems to be asking more than anyone could deliver.

That said, if the fact that there is anything at all can't be satisfactorily explained in any of the ways we explain other things, that's interesting. If the question just draws us toward that conclusion, it's worth asking.

Lastly, one thing that explanations do is provide people peace of mind or the feeling of understanding. We ask for explanations when we feel confused or lost. The question might just be a request for *something* that gives you the feeling of understanding the fact that stuff exists. In that case, the question doesn't have a single correct answer: what gives your neighbor the feeling of understanding might not do

---

[*]   You don't actually need to know how Peano arithmetic works in order to get the point here, but see page 143 for the details.

the same for you. And of course, I can't tell you what's going to give you the feeling of understanding here, because I don't know who's reading this. You'll have to find it on your own.

This discussion has conspicuously left out any mention of God, who's often invoked in this context. I don't think God will help us answer the question, in part for the reasons I gave above. But there's another big reason I'm leaving God out of the picture. . . .

Does God Exist?

God (or a god, if you're not into the whole monotheism thing) is an all-powerful, all-knowing being who wants the best for the world. If such a being existed, the world would be perfect. The world is not perfect. So, there is no God. ☹☹☹

> ✎ I agree that the world isn't perfect. But who said that God is an all-powerful, all-knowing being who wants the best for the world? This conception of God seems weirdly tailored to making this argument work. In any case, that's not my God.

Fair! Different people have different things in mind when they talk about God, which is part of the problem.

There's a way around this, though. Whatever (most?) people have in mind when they're talking about God, they're talking about something that it makes sense to worship. So what is worship? To worship something is, more or less, to submit your will to it completely, because you value and trust it that much more highly than you value and trust yourself and your own judgment.*

---

* This isn't obvious. Worshiping something is a way of valuing it very highly, clearly, but why bring submission into it? Well, you need to take into account, somehow, that worship isn't a relation between equals. No matter how much I love my partner, it's only metaphor or hyperbole to say that I worship her. The idea of submission seems to me to capture the essential and immutable hierarchy that's built into the idea of worship.

But does it ever make sense to submit your will to something in this way? It's kind of undignified, for starters. You also run the all-too-real risk of submitting your will to the wrong thing. There are, let's say, a couple of historical examples of people putting their money on the wrong God. And even if God *is* perfectly trustworthy, *you* are not, so you shouldn't be so confident in your own judgments of God's trustworthiness that you totally hand over the keys. So it doesn't make sense to worship anything. So, God doesn't exist.

Silver lining: you can still go to church/mosque/temple/shul/wherever and participate in meaningful religious rituals. The question whether God exists is not the question whether your religious *practices* are a mistake. Community, storytelling, pageantry, the holidays, with their ritual food and music—all of these things retain their value in the absence of God. After all, Christmas is still fun even after you find out (spoiler alert) that there is no Santa.

> A teenager came to the booth accompanied by his mom, who clearly did not want to be there. He asked whether God exists. I gave a one or two sentence version of this answer. The teenager grinned, and the mom let out something between a gasp and a wail. Maybe I shouldn't have enjoyed this as much as I did.

What Is the Meaning of Life?

I don't know, but it doesn't matter.

Suppose you found out that the creationists are right—that human beings were put on earth by a group of alien livestock farmers. They wanted us to populate the earth, so that as soon as possible they could come back and eat us. The faster we breed and the tastier we become, the better.

If this story were true, your life and human life as a whole would have a clear purpose: you were meant to help feed aliens. But so what? It wouldn't be especially comforting or helpful to find out that you were meant to help feed aliens. Even if you were meant to help feed aliens, that doesn't mean that you *should* help feed aliens. (If anything, you should work to *prevent* the aliens from eating us.) The point is that even if your life did have a meaning, discovering that meaning wouldn't have the emotional and practical significance that we typically take it to have.

So what *would* have that emotional and practical significance? People tend to think more about the meaning of life when they feel dissatisfied with how their lives are going or suspect they were wrong to pursue the sorts of careers or life projects they've pursued. The meaning of life, whatever it is, is supposed to offer guidance to people facing these sorts of situations. So where can you get this guidance? One place to look is at the psychological research on job satisfaction, at least some of which suggests that people tend to be more satisfied with their jobs when they get to interact with other people, they have a considerable degree of autonomy, they use that autonomy to exercise special skills, and they believe they work for an organization that makes the world a better place. But that might not be enough for you. If you think your work does a little bit of good for the world, but you want to do a *lot* of good, (a) that's very nice and (b) congratulations! You're an effective altruist. Go google it.

Do We Have Free Will?

This is a funny question. Usually, if a philosophical problem keeps a lot of (otherwise) ordinary people up at night, it's a problem that can be stated in plain language.* But the problem of free will revolves around a piece of technical jargon—namely, the phrase "free will." How we answer the question depends in large part on how we choose to define "free will." Here are some possible definitions and the answers they lead to.

> DEFINITION #1
> Free will is the ability to make choices.

Yes, we have free will because we make choices all the time. Suppose a schmoozy mom goes to the supermarket. While she is lost in conversation with a fellow shopper, a jar of peanut butter falls into her cart, and she pays for it at checkout without noticing. She doesn't choose the particular brand of peanut butter she purchases. Moments later, a choosy mom enters the same supermarket. She looks over all the types of peanut butter on sale and compares them with respect to price, ingredient list, and whatever else. She picks one up, puts it in

---

* Although the more I think about this, the more wrong it seems. People come to the Ask a Philosopher booth with a *lot* of jargon-y or otherwise scholarly questions—about subjectivity, objectivity, dualism, normativity, etc. Who knows why this is? Maybe it's a symptom of a deeper problem—that people think of talking about philosophy as talking about what some philosophers have thought.

her cart, and, quick as a jiff, proceeds to checkout. The choosy mom *does* choose her brand of peanut butter.

There's an important difference between the schmoozy mom and the choosy mom, which we mark by saying that only the latter makes a choice. If free will is just the ability to make choices, denying that we have free will would just mean that there's no such difference between schmoozy mom and choosy mom. That would be silly.

✍ OK, but what is the fundamental difference between voluntary and involuntary actions?

I have no idea. Is it a matter of having a feeling of control? Is it a matter of the reasoning or information processing that precipitated the action? (What sort of reasoning or information processing, then?) Is it a matter of whether you perform the action consciously? Is it a matter of wanting to perform the action deep down? (Deep down where?) Is it some combination of these or something else altogether? All of those answers are plausible enough that it seems wise not to have an opinion.

> **DEFINITION #2**
> If you act on your own free will, you could have acted differently.

Yes again. Maybe it will turn out that our commonsense beliefs about what could have been are somehow radically mistaken. (Why *do* we have commonsense beliefs about how things could have been, after all? Why do we care about how things could have been as opposed to how they actually are?) But if we can take that common sense for granted, there are all sorts of situations in which we could have acted differently. This morning, I put on a gray shirt, but I could have put on a black shirt or a red shirt or any other shirt. Our actions aren't special in this respect. It rained yesterday but not so bad; it could have been worse. I just flipped a coin and it came up tails; it could have come up heads.

If we could never have acted differently, either our actions are somehow determined in a way that the rain and coin tosses and other natural phenomena aren't, or we are wrong across the board about how things could have been. Why think that?

> **DEFINITION #3**
> Free will is the ability to act in a way that is not determined by the laws of physics applied to one's body and environment.

No, we don't have free will, because our bodies obey the laws of physics. But also, why would anyone *want* her body to violate the laws of physics? (I get why people want to fly or whatever, but this is not that.)

✎ But something-something quantum randomness!

Yes, perhaps we live in a world where what actually happens isn't necessitated by the laws of nature. But even if the laws of nature only yield probabilities—not determinate outcomes for this or that situation—your actions relate to the laws of physics in roughly the same way as the "actions" of an inanimate object.

> **DEFINITION #4**
> Free will is whatever makes it appropriate for people to hold you morally responsible for your actions.

Yes, we have free will because it is sometimes appropriate to hold people morally responsible for their actions. We blame people, we praise people, we hold people to account all day long. It would be a remarkable coincidence if this practice didn't at least sometimes, on balance, work out for the best. It does seem to me, though, that the more we think of our behavior as physically or biologically caused, the more likely we'll be to treat bad behavior as an engineering problem. Instead of holding people responsible, we'll educate and persuade and medicate them to do better. If we ever reach a point where bad behavior becomes *nothing but* an engineering problem, we'd lose our free will. Would that be so bad?

How Do We Know Anything About the World Outside of Our Own Heads?

What makes you think you know anything about the world *inside* your own head? Imagine, as vividly as you can, the view of your childhood home from across the street. How much of the scene is included in the image? How much color does it have? How detailed is it? If you're like me, you'll find these questions incredibly hard to answer. This might begin to shake your confidence that you know your own inner life better than you know prosaic facts about the world around you and that knowing the external world is a matter of piercing the veil of perception.

But suppose that there are some things that you can know with certainty about your experiences right now—say, that you seem to see some squiggles on the page in front of you—and we want to know how you can get from knowledge of this experience to knowledge of the external world. This experience, we believe, is typically caused by light reflecting off of a page, hitting the surface of your eyes, which then send some electrical signals through your nervous system. But couldn't this experience be caused in some other way? Couldn't you be hallucinating? Couldn't you be dreaming that you're looking at a book? Couldn't a mad scientist directly stimulate your brain in just the right way to cause the experiences you're having?

Well, maybe. You can't conclusively rule these possibilities out, in the sense that they are all logically consistent with you having the exact stream of experience that you have had in your life up to this point.

But even if you can't conclusively rule these possibilities out, starting from the evidence of your senses, you can still be reasonably certain that a broad chunk of your commonsense beliefs about the world are right. The answer, in a nutshell, is abduction. (That's *ab*duction, not ab*duc*tion, which is almost never the answer.) Deduction is the sort of reasoning that you study in a typical logic class: Socrates is a man; all men are mortal; so Socrates is mortal. The conclusion of a deductive inference is no more informative than the premises. Induction, on the other hand, is the sort of reasoning that you study in a typical statistics or probability class: the average height of all the adults in this random sample is five feet seven inches; so the average height of all adults everywhere is five feet seven inches, give or take. The conclusion of an inductive inference can be more informative than the premises, but the conclusion is stated in the same vocabulary as the premises: if the evidence you bring to the table is about how tall people are, the conclusion will also be about how tall people are.

Abduction—or inference to the best explanation—is different. When I was a kid, I moved into my oldest brother's old bedroom. As I was settling in, I looked in the closet and saw I HATE AIN scrawled on the wall in red crayon. Now, it's *possible* that my mom wrote this. But my mom doesn't hate me, she knows how to spell my first name, this isn't her handwriting, and even in the nineties, she was a little too old to be writing in crayon. A much better explanation of how those words got there is just that my brother scribbled them in a fit of anger. You can think of other examples: inferring that it rained overnight when you see a wet sidewalk in the morning, inferring that someone is trying to get your attention when they call your name, inferring that Venus orbits the sun more closely than we do from the fact that it waxes and wanes. Unlike induction and deduction, abduction allows you to draw an inference from an observation couched in one vocabulary to an explanation of that observation couched in a completely different vocabulary.

However we describe the evidence that we get from our perceptual experiences, including the evidence from other people telling us about what *they* see and think, the best explanation of that evidence will appeal to more or less stable objects that exist independently of whether I am perceiving them at one moment or another. Why, for example, do I seem to see my laptop screen in front of me? Why do exactly the words that I seem to type appear on that screen? Why, when I seem to (don't try this at home) ask the person sitting next to me if there's really a computer on my lap, do they seem to say yes? We *could* explain all of these things in terms of hallucinations, dreams, or mad neuroscientists, but these explanations aren't as good as the natural one—namely, that I'm sitting at my laptop, typing these very words.

✍ But what makes one explanation "better" than another?

Who knows? It will, however, have something to do with simplicity, ad hoc-ness, how much we can explain on the same basic pattern, and how much each explanation requires us to revise our previously held beliefs. The commonsense explanation beats the far-out skeptical explanations on all four counts.

✍ Wouldn't a hard-core skeptic, someone who believed that no one knows or has any reason to believe anything, find this argument unconvincing? After all, they'd just say, "Why believe in abduction?"

There's a lot to say about this, but one thing that's helpful to remember is that the question "How do I know anything about the world outside my own head?" isn't the same as the question "What would it take to convince a hard-core skeptic that I know this stuff?" or even "How do I know that abduction works?" Abduction might be a good answer to the first question, even if it's not a good answer to the other two.

A girl who looked about five years old passed by with her mother. The mother asked if she had any questions. "How am I real?" One of the philosophers answered, "Close your eyes. Are you still there? Then you're real." The mother quickly whisked the girl away. The daughter scratched her forehead on her walk toward the train.

If the World Is Melting, Is It
OK to Have Kids?

This is tough. To really do it justice, we'd have to answer some deep, messy empirical and evaluative questions about what life will be like on a warmer planet, the average person's lifetime contribution to climate change, what parents owe their children, why people want children in the first place, the morally significant differences between possible future people and people who already exist, the ethics of adoption, how individual responsibility figures in to collective action problems, and where we draw the line between permissible but non-ideal actions and totally impermissible actions. I won't attempt to answer those questions.

But: one big reason why global warming is supposed to raise some scruples about reproduction is that life on a warmer planet is likely to be much worse than the life that we enjoy now. What if it's so bad that it's not worth living at all?

Well, how do you tell whether a life is good enough to be worth living? The easiest answer is just to ask living people whether they think their lives are worth living. It's notable that most people who say no (perhaps excluding people with painful, debilitating terminal illnesses, who have a firm grasp on their life prospects) are suffering from a mental illness that systematically distorts how they value things. (This might seem circular, but it's not. We have ways of diagnosing mental illness other than asking people whether they think their own lives are worth living.) But maybe we're all biased in this regard—by selective memory, by our fear of death, by the difficulty of imagining lives all that different from our own. One way around these potential biases is just to ask of any given day whether it was bad enough that people wish they had never lived it or had just slept through it. Or again, to take a page from what psychologists call experience sampling, we could text a group of people at random points throughout the day and ask them to rate how they're doing at that particular moment on a scale from one to seven, perhaps along a few different dimensions (enthusiasm, contentment, relaxation, etc.). I don't know, but my guess is that the percentage of people who would say that they were having *such* a bad day or whose average response to the text was below a four will be quite small—less than 5 percent, say.

Of course, this is all *before* the worst effects of global warming come to pass. But then again, no matter how bad global warming is, will the life of the average person after global warming be much worse than the worst 5 percent of lives today? Maybe, but as long as these lives aren't that bad, they will still be worth living.

In any case, everyone who wants a child should just adopt one.[*]

> One visitor said she wasn't surprised to read about recent research on the moral beliefs of chimps because she grew up on a farm where she saw that all sorts of farmed animals have moral beliefs. I asked, "What's the dumbest animal that has moral beliefs?" She said, "Human beings."

---

[*] Don't @ me.

How Do Brains Give Rise to
Conscious Experience?

No one knows. One way that you know that no one knows is that the different theories of how brains give rise to conscious experiences entail very different conclusions about how many creatures in the world have conscious experiences. We have no way of settling which theory is right on this count that doesn't assume one of these theories in the background. So, at least for now, we have no way of answering the question.

We can make progress on it, though. One way we can make progress is to get better at describing our experiences themselves. We might test for descriptive precision by, for example, measuring test-retest reliability, seeing what sorts of people in what sorts of situations are more or less likely to fall prey to what psychologists call inattentional blindness, or assessing the extent to which people are able to bracket their knowledge of the outside world (say, in the construction of subtle visual illusions) in describing their experiences.

Another way to make progress on the question is to be as clear as possible about what we mean by "consciousness." People sometimes use "consciousness" to talk about knowledge of morally or politically important problems, but that's evidently not what people use the word to mean in this context. We can distinguish—at least—creature, state, and qualitative consciousness. We call a creature conscious when it's fully awake rather than asleep or comatose. We call a mental state—a fear, hope, belief, or itch, say—conscious, roughly, when we know that we have it but not as the result of an inference. You have qualitative consciousness, on the other hand, if there is something it is like to be you or if you have an inner life or stream of experiences. These three types of consciousness can come apart. For example, the nematode *C. elegans*, with all of its three hundred–some neurons, has a sort of sleep cycle. It is capable of creature consciousness, then, but I doubt it has any conscious mental states or an inner life. Or again, a feeling of irritation might creep into your stream of experience, but you don't become conscious of this felt experience until someone turns off a noisy air conditioner nearby and you find yourself suddenly relaxed. So not all of our inner life is state conscious.

The problem of how brains give rise to creature and state consciousness is hard, but we know it's not impossible. After all, we have more or less accepted behavioral and physiological tests for these things, at least among typical human adults. The problem of how brains give rise to qualitative consciousness is harder—in part because we have no such tests for people's inner lives. But if the history of science shows us anything, it's that declaring for philosophical reasons that such and such is impossible to explain scientifically is a good way to make an ass out of yourself once some enterprising young scientists prove you wrong.

Why Should I Care?

If you're looking for a proof from incontrovertible factual premises that you should care about something or other, you won't find it. But good news! You don't have a choice whether to care. If you ever feel angry, sad, happy, or proud of anything, you care about it. Since it's not in your power never to feel these things, it's not in your power never to care. The question is whether you care as well as you can—in a consistent way that stands up to informed reflection. Maybe you used to care whether your team wins the big game, but the more you learn about the sport, the less connected you feel to your team or to the sport as a whole. Maybe you didn't care about politics, but the more you read about current events and history, the more strongly you feel. Reasoning alone can't get a completely apathetic person to care. But luckily, it doesn't have to, because no one, more or less, is completely apathetic. It can get you to care in a more consistent, reflective way, though, and that's all we can ask for.

> ✎ Isn't this disappointing, though? If everything I care about is basically what I'm forced to care about by the circumstances I was born into, doesn't that make caring seem sort of groundless?

We *could* be upset by the fact that you can't prove that something is worth caring about from purely factual premises. But the facts themselves don't rationally compel you to be upset.

For my part, I am from time to time overcome with panic and terror by certain features of the world that are completely beyond my control—my own mortality, the mortality of everyone I love, the coming supernova of our own sun, the growing chasm between each galaxy and the next, the (so I'm told) inevitable heat death of the universe. I can't easily shake these feelings, and I doubt I'll ever fully be able to. But when I'm in their grip, it seems like they're the *correct* response to the facts at hand, like failing to feel these things is tantamount to failing to see what's right in front of my face. It's helpful, even therapeutic, to remind myself that this is a mistake. While emotions can be reasonable or unreasonable, they aren't the sorts of things that can be true or false. If you're one of those people who really cares about avoiding what's false, this can be liberating.

What's the Best Form of Government?

L ike most philosophical questions, this is a bit vague. There are lots of different properties that a government can have that you might call forms, and many of them are either compatible with one another or have nothing to do with one another. But maybe this will count as a satisfying answer.

I'm a socialist. I take socialism to be the view that we should work toward the collective and democratic ownership and management of many things that are currently private property. That leaves a lot unanswered: "Work toward" by what means? (The labor movement, electoral politics, violent political revolution?) Which collectives? (Workers, subjects of a municipal, state, federal, or world government?) Democratic in what sense? (Representative democracy, something more participatory?) Whose private property? (Individual businesses, industries, entire economies?). But this isn't the place to fill out these details.

Here are the arguments for socialism that I find most convincing:

THE ARGUMENT FROM INEQUALITY

Capitalism—the view that goods and services should be provisioned primarily by private entities interacting through markets—leads to massive inequality. You might think this is intrinsically bad because it is unfair. But I think of the badness of inequality in terms of waste and domination. Inequality leads to waste because resources that would be better off in the hands of the poor are put in the hands of people who don't need them; inequality leads to domination because when some people are much richer than others, they can manipulate the basic institutions of society so that things keep going their way.

Here's another way to think about it. To a first approximation, when a society provides a good or service via the market, that is its way of saying that it's OK if poor people don't have access to it.* Of course, we've already decided that there are some goods that it is not OK to deny poor people—roads, the postal service, K–12 education, Social Security, libraries. But why stop there? Are we really OK if poor people don't have food, shelter, clothing, medical care, or access to toilets, the internet, or higher education? Not me.

THE ARGUMENT FROM GLOBAL WARMING

Capitalism depends on constant growth, and, at least as a rule, constant growth means constant increase in the use of energy and natural resources. This is unsustainable.

Not every form of socialism is environmentally sustainable; look at Venezuela or Norway. But *some* form of political-economic system that isn't based on perpetual growth is necessary if we are going to avoid the existential risk that climate change poses.

THE LIMITS OF MARKETS

Since Adam Smith, classical economists have used the idea of the invisible hand to describe the ways in which individual people acting in their own self-interest can, through their interactions in a market, benefit each other. Consider the iPhone. Apple is primarily interested in making money. One way they can make money is by continually developing fun and useful new features for their phones and selling

---

* Yes, the United States has a mixed economy, where people who can't afford certain goods and services get them through means-tested government subsidies or public services like public housing. But means-tested services are always precarious and underfunded because they're always stigmatized. Compare food stamps with Social Security.

their phones at prices that won't scare too many consumers away. Individual consumers buying iPhones are also acting primarily in their own self-interest, but the money that they pay Apple funds further development. Everyone wins! (At least as long as we don't look too hard at the working conditions in factories that manufacture iPhone parts or the environmental costs of mining for the rare minerals used in iPhones.) Defenses of free markets often appeal to the power of the invisible hand.

But when you think about the range of conditions or types of markets under which the invisible hand fails to work—when we're dealing with monopolies and cartels, unequal information between buyer and seller, tragedies of the commons, other negative externalities, public goods, planned obsolescence, rent seeking, cronyism, markets that manufacture the preferences they satisfy—it becomes more plausible that this is a system we should, at least in large part, replace, not a system we should fix.

THE ARGUMENT FROM AUTOMATION

Something incredible has happened! We now have robots and machines to do a lot of the mindless jobs no one wanted to do before. But people are terrified of this. This might be a familiar point, but we should really reflect on how bonkers this is. And it's all because we have privatized the benefits of automation.

THE ARGUMENT FROM LABOR ALLOCATION

Anyone who has worked at a medium- to large-size business knows that lots of jobs shouldn't exist. To some extent, this is because managers solve problems by hiring new managers. To some extent, it's because once you create a job, it can be hard to get rid of. But as the anthropologist David Graeber has shown, an enormous number of people think that *their own jobs* shouldn't exist—telemarketers, debt traders and collectors, secretaries and assistants for people who don't need secretaries or assistants, and so on.

By contrast, think seriously about the most important jobs that you could possibly do. Here are some candidates: researching clean meat, pandemic research, existential risk research, carbon capture and renewable energy research, low-cost overseas public health interventions. Most of the funding for most of these jobs comes from private charity and public grants. Or think of the difference in funding for research, treatment, and prevention of diseases of poverty like malaria versus diseases of affluence like hypertension. That is because they are not in the short term profitable or because they run contrary to the interests of the fabulously wealthy. A better political and economic system, in which profit and the interests of the wealthy mattered less, would fund these things more generously. (We could make a similar but slightly different argument about vital but wildly underpaid professions—social workers, home attendants, janitors, farmworkers, warehouse workers, childcare providers, etc.)

THE ARGUMENT FROM MORAL SHIFTINESS

Many of the early supporters of free markets believed in them because they thought they would lead to a more equal society. This was plausible because (a) free markets challenged many of the greatest sources of inequality at the time (feudal property laws and guild monopolies, for example) and (b) these early supporters generally thought of each worker as self-employed or on the road to self-employment: this was before the Industrial Revolution.

Capitalism has failed spectacularly at accomplishing those goals. But instead of working toward a better form of political economy, we have changed the underlying morality of capitalism. Now it's about what we personally deserve, meritocracy, and freedom from government interference. This should at least give us pause.

INTERSECTIONS WITH OTHER SOCIAL ISSUES

To a hard to specify extent, capitalism exacerbates social problems that aren't entirely economic in nature—factory farming, war, ableism, racism, police brutality, sexual violence, the exclusion of women from various dimensions of public life. Socialism is not the complete solution to these problems, but it is one key part of the solution.

> The boy who asked the question about the best form of government had been at the booth for about fifteen minutes with two younger friends. They had rifled through the bowls of questions and thought experiments, speed talking with us about this and that. Then one of the philosophers asked the boys if they had any questions of their own. The oldest boy screwed up his face and suddenly became a little sheepish. He paused for a little bit. "I'm not sure if this is a philosophy question." "That's OK. Try it out." "What's the best form of government?" For the record, this is the fundamental question of political philosophy, and yes, it is a philosophical question.

Is Color Subjective?

S ubjective" is a slippery fish. There are at least two things it can mean. The first has to do with knowledge and reasoning. The second has to do with what makes something what it is. Color is subjective in the first sense if maximally reasonable, well-informed people could disagree about the colors of things. Color is subjective in the second sense if nothing has a color full stop—if things only have colors relative to an individual person. That is, color is subjective in the second sense if color is like the time of day or being to the left of something. It is never simply 11:11 a.m.; it is only 11:11 a.m. *in* this or that time zone. Nothing is ever simply to the left of something; it is only to the left *relative* to a point of view. If color is subjective in this sense, when we say that something is red or pink or orange or whatever, we're really, in part, talking about ourselves.

(Digression: It is very easy to mix up the first sort of subjectivity, which philosophers sometimes call epistemic [from the Greek word for knowledge], with the second, which we sometimes call metaphysical [because it's about the way the world really is, not just how we think or talk about it]. Starting from the observation that reasonable people disagree about something and inferring that they're really just talking about themselves is a quick way to get very confused.)

At the risk of dating myself, let's talk about the Dress. If you're reading this after this meme has expired (RIP): it's a weirdly lit picture of a Dress that you can see as either blue and black or white and gold. Some people can make the switch between the two ways of seeing the Dress, but most people can't. Eventually, the world learned that the Dress was *really* blue and black—that is, that people who saw the Dress in real life across a range of lighting conditions agreed that it was blue and black. But even after this news came out, some people insisted that the Dress was white and gold. When you know that the Dress is really blue and black but insist that it's white and gold, what you are insisting, roughly, is that it looks white and gold *to you*. But then you don't really disagree with the people who say that it's blue and black. It's a bit like two people sitting across a table from each other, one of whom says that the salad is to the left of the pasta, the other of whom says that the salad is to the right—not the left—of the pasta. They don't disagree, because they're each talking about their own perspective. Sometimes, when we're talking about color, we're talking about something metaphysically subjective.

But not all the time. Before we found out that the Dress was really blue and black, a lot of people who said that the Dress was white and gold weren't just talking about how this particular picture looked to them. They believed that if you looked at the Dress in a range of lighting conditions, you would conclude that it was white and gold. They were wrong, and they changed their minds when they found out that the Dress was really blue and black. Sometimes, when we're talking about color, we're not talking about ourselves; there's nothing metaphysically subjective about it.

    🖋 Yeah, but even if the Dress really is blue and black, what *makes* it blue and black is how people experience it. Things really do have colors, but what gives them colors, at least in part, is that we experience them in a certain way. If no one ever saw color, nothing would be blue or black or white or gold or anything. Color exists only in or because of our heads. That's what I mean when I say that color is subjective.

Maybe, but why think this? It's helpful to compare color to shape. The apparent shape of a thing can vary with your perspective and the lighting conditions. Your perception of shape, like your perception of color, is subject to all sorts of distortions. (Personal favorite: the convex face illusion, in which a face may seem like it's in relief like the front of a mask or impressed like the back of a mask.) But while we (at least today) are quick to jump to the conclusion that color is just in the head, I don't know of many people who think the same about shape.

There's at least one difference between color and shape that might be relevant here. Modern physics describes and explains all sorts of things in terms of shapes. But think about the weird set of things that we call red—red wine (which is the color of purple grape juice), red watermelons (which are red on the inside), red apples (which are red on the outside), red patches of color in our visual fields (which aren't distal physical objects at all), red daubs of paint (which we would call orange or brown if we saw them in different contexts), and so on. These things have nothing physically in common. Physicists might explain the behavior of an object in terms of the frequencies of light that it reflects, but they won't explain it in terms of its redness or greenness or whatever. They can get by just fine without ever talking about the colors of things.

Then again, physicists can also get by just fine without talking about chairs. That doesn't mean that chairs exist only in our heads. Right?

Is Time Travel Possible?

I t's not just possible; it's real. Set two very accurate clocks to the same time. Leave one on earth, and take the other up in an airplane for a little while. When the plane lands and you compare the two, the plane clock will be a tiny bit earlier than the earth clock. In principle, if the plane got close to the speed of light, the plane clock would be a *lot* earlier than the earth clock. If a person spent, say, a year on that plane, many years might have passed on earth. (Physicists call this time dilation.)

But while this is incredibly cool, it's likely a bit disappointing. After all, this is *forward* time travel. The only difference between this and the type of time travel we're all doing just by being alive is that the person on the plane is traveling through time noticeably more slowly than usual. But most (though not all!) of our time travel fantasies involve traveling *backward* through time. Is this possible?

Well, there's a well-known argument that it's not—not just according to physical theory or given our technological limitations but in principle. This is the so-called grandfather paradox. If I could travel back in time, I could kill my grandfather. But I couldn't kill my grandfather, because if I killed my grandfather, I would never have been born, and so I *wouldn't* have killed my grandfather. So I couldn't travel back in time.

Maybe, but this argument makes some assumptions that as far as I can tell might not be true. Maybe I could travel back in time, but only if I don't do anything that leads to paradox. Maybe time has a branching structure, so that when I travel back in time and kill my grandfather, I prevent the version of me *on that branch* from being born, not the version of me that actually did the killing. Maybe time has a looping structure, so that when I travel back in time and kill my grandfather, I prevent the loop on which I was born from happening *again*, but it already happened once. Or something.

But there's a nearby question that came up at the Ask a Philosopher booth. The guy who brought the topic up was thinking of backward time travel. We talked for a bit about the grandfather paradox, inconclusively. Then I asked him why he wanted to travel backward in time in the first place. He gave some sort of familiar reasons—correcting mistakes, reliving experiences that he couldn't appreciate the first time around. But what about forward time travel? What practical point would that serve? His response was really thought provoking: someone who traveled forward in time would care less than other people about the ephemeral events of the day. The greater the span of time you have personally experienced, the broader your perspective, the less invested you are in each individual moment. (After all, doesn't something like that happen with aging in the non-time-dilated world? Or is this a bad description of what aging does to your relationship to time?) Such a person, this guy suggested, would be especially useful in our political decision making.

There's a surprisingly plausible argument for gerontocracy here. Even if we didn't or couldn't answer this guy's questions about the possibility of backward time travel, it pays to follow the thread out. Philosophical questions have a way of making connections to each other.

Are People Innately
Good or Bad?

A little bit of both. People are born predisposed to do all sorts of good things. The philosopher Mengzi asks us to imagine seeing a baby teetering at the edge of a well, about to fall in. Most people he thinks would run to protect the baby without giving it a thought. If you wouldn't run to protect the baby, you probably had to *learn* not to (unless you're some kind of born psychopath). You can think of your own examples of more or less universal (and therefore probably innate) good tendencies that people have—to console crying people, to care for our offspring, to cooperate in all of the ways that make communal life possible.

People are also born predisposed to do all sorts of bad things. We form more or less arbitrary groups—think of the infinitesimal doctrinal disagreements between different sects of Christianity—and grow cold or cruel to the people who fall outside of them. We care more about attractive people, people who look like us, people who are close to us in space and time.* (Environmental policy would look a little different if the kids who will be born in thirty years had a vote I imagine.) We are unmoved by statistical descriptions of large-scale misery. We are biased in favor of existing (and often unjust) social structures and conventions. Plato asks us to imagine what you would do if you had the mythical ring of Gyges, which makes its wearer invisible. If you're being honest, I think the answer will include some sneaky stuff—which is to say that, if you're being honest, there's some sneaky stuff you don't do only because you're afraid you couldn't get away with it.

I'm interpreting this as a question about people's innate predispositions, though—the sorts of things we'll do, other things being equal. But you might interpret it as a question about whether people are born and *stay* good or bad. You might also interpret it as a question about human nature (whatever that's supposed to mean) or how people are deep down. I'm not sure if these other questions are more important somehow. (Why do we care whether people are *innately* good or bad, anyway?)

It's also worth reflecting on why our innate predispositions are sometimes such a bad fit for modern life. In part, it's because they evolved among hunter-gatherers who lived in small groups under conditions of extreme scarcity—people who lived very differently than us. Preferring attractive people and people who look like us might have helped keep our prehistoric ancestors around, but today

---

* "But," you protest, "isn't it totally OK to care about nearby people more than people overseas?" The philosopher Peter Singer famously asks us to imagine passing a child drowning in a shallow pond. You could easily wade in and save the child from drowning, but you would ruin your clothes in the process. If you refused to help the kid because you were worried about your own clothes, people would think you were a monster. But why is that? Well, one natural explanation is that if you can make something really good happen (e.g., saving a child's life) without sacrificing anything of comparable importance (e.g., your clothes), you should. But how differently would we relate to people in other parts of the world if we followed this principle?

it just makes you a jerk. Whether people are innately good or bad is to some extent a matter of how well our ancestral environments match the environments we live in now. To that extent, we would expect ourselves to become innately worse the further modern life departs from the past. On the bright side, you have something to look forward to if we ever go back to the Stone Age.

One of the philosophers asked a visitor about the ring of Gyges. The visitor answered: Aren't blind people surrounded by invisible people all the time? Don't they manage not to be treated awfully? I can't get over how incisive and fruitful this response is. You can reason counterfactually about the effects of invisibility on people's behavior, but you don't have to—just look at the various ways that our behavior is visible or invisible in the real world.

Which Came First,
Thought or Language?

Thought.

One way to find out that prelinguistic babies have thoughts—that is, that they believe things—is to look at them when they act surprised. In one classic study (illustrated in the figure below),* babies face the wide part of a drawbridge. They see the drawbridge go up and down a few times, rotating a full 180°, until they get used to it. Then, when the drawbridge is lying flat facing the baby, the experimenters place a block in the path of the drawbridge.

The Drawbridge Experiment

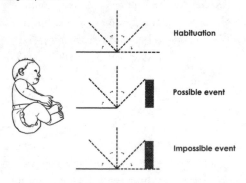

Habituation

Possible event

Impossible event

---

* Renee Baillargeon, Elizabeth Spelke, and Stanley Wasserman, "Object Permanence in Five-Month-Old Infants," *Cognition* 20, no. 3 (1985).

One group of babies then sees a "possible event": the drawbridge rotates until it occludes the block. When it reaches the block (which the baby can't see), it stops and rotates back toward the baby. After all, the drawbridge can't go *through* the block.

Another group of babies sees an "impossible event": the drawbridge rotates until it occludes the block. Then the experimenters remove the block (out of sight of the baby), and the drawbridge rotates the full 180°. The block has vanished! Even weirder, as the drawbridge rotates back, the experimenters put the block back in place. The block has magically reappeared!

Here's the crucial thing: the babies stare at the impossible event several seconds longer on average than they stare at the possible event. The most natural interpretation of this result is that they expected the block to continue to exist even when it was occluded by the drawbridge, and that expectation didn't pan out. This interpretation is supported by variations on this experiment, using other impossible events, other measures of surprise (heart rate, sucking, smiling), and infants as young as two and a half months old (!). Anyway, expectations are a type of belief, so babies who can't speak or understand any human language have beliefs.

You could make the same point with our nonhuman friends. (They have more or less ornate signaling systems, but they don't have languages that come close to human language in expressive power or grammatical complexity.) My dog* has all sorts of beliefs. For example, if my partner leaves to take him for a walk, and I'm not there when he gets back, he runs around the apartment seemingly looking for me. The most natural explanation of this behavior is that he expects me to be home when he returns. He has other expectations revealed in similar ways: expectations to do with what happens when you say "treats," where his food is located, what's in store when you walk toward the park. He's a good boy, but in this respect he's not an exceptional boy.

---

* Instagram: @donteatscrapple.

But are these really beliefs? Couldn't they just be patterns of behavior or something?

The short answer is that positing baby and dog beliefs enables us to predict and explain a bunch of baby and dog behavior accurately and succinctly. So my belief in baby and dog beliefs is on the same footing as my belief in anything that enables us to predict and explain a bunch of things—a pretty good one.

But even if babies and dogs can have *some* beliefs without language, their beliefs tend to be about sort of boring things—mostly stuff that they can directly perceive. Grown-up beliefs are different. They stretch into the distant past and beyond what we see and touch. For example, I believe that Plato wrote *The Republic*, that electrons have a negative charge, and that the cube root of 27 is 3. It's hard (though perhaps not impossible) to imagine how anyone could acquire any of those beliefs without language. Language enables us to refer to people and subatomic particles and mathematical functions and other things that we can't perceive, which is otherwise very difficult to do.

I think this is really profound. In philosophy, we seem to run up against a wall from time to time—the limits of what we can think about. When have we really hit the wall, and when are we just temporarily failing to think straight? I don't know, but the answer will have something to do with how language can—and can't—enable us to think new thoughts.

How Well Do We Know
What's Good for Us?

There are certain ways in which we are all experts on ourselves. I am one of the few living people who know my Social Security number, how much money I have in the bank, whether I'm hungry right now, what I ate for breakfast, my plan for the day, and so on. With the minor exception of Google and everyone they sell my personal information to, I am the world's leading authority on my location just about all of the time. This information isn't completely idle. It is crucial to figuring out what I should do for lunch, among other things. So in some respects, we are experts in what's good for us. We are also, in certain respects, experts in what's good for people *like* us. If you exclude certain demographic groups from certain sorts of conversations, you'll miss out on the concerns and insights that those groups bring to the table. For example, one reason medical conditions that present differently in women are especially likely to be misdiagnosed is that medical research has historically been performed by men, often on male subjects. At least some of the time, if you want to know what's good for a demographic group (or what's good for a larger group to which they belong), you should ask members of that demographic group.

On the other hand, psychologists and behavioral economists have, in the past few decades, produced mountains of evidence that our sense of what is good or bad for us is sensitive to all sorts of irrelevant distractions. We buy things when we're told they're on sale that we wouldn't buy at full price, even when the price is the same; we make identical choices differently when they're framed in terms of risks or rewards; we do things we don't want to do just because we've already paid for them; we spend money with credit cards that we would never spend in cash. And again, sometimes we are systematically mistaken not only about what's good for ourselves, but also about what's good for people who share our social positions. Karl Marx called this false consciousness. Perhaps medieval peasants believed that the meek shall inherit the earth because it's really true; but more likely they believed it because it was very convenient for their lords to keep that belief going strong. History has played host to a shocking number of black defenders of African colonialism and the transatlantic slave trade. As Thomas Frank argues in *What's the Matter with Kansas?*, voters are routinely distracted from their economic interests by (often more emotionally affecting) culture war issues. In general, people are sometimes very wrong about how well the prevailing social arrangement is working out for their groups, especially when their ignorance is in the interest of the powerful.

Where does that leave us? It's not terribly helpful to say, "Sometimes we know a lot about what's good for us, and sometimes we don't." For a more helpful, general account of how what seems good for us differs from what is good for us, you should probably check out an introduction to behavioral economics. But what about how well we know what's good for people who share our social positions? When will people in a social group have distinctive, relevant insights into their own condition that outsiders lack, and when will they be trapped in false consciousness? I don't have any kind of general answer to this question, but certainly *seeking out* those insights when you can is always a good policy. If you don't bring members of some social group into a discussion of what would be good or bad for them (or what would be good or bad for larger groups of which they're a part), you're going to miss whatever insights they might have.

One of the visitors to the booth was an adult with intellectual disabilities on a walk with her caretaker. She dug around the bowl of philosophical questions we set out and picked: "What was the happiest day of your life?" We talked about some big celebrations, the connection between happiness and specialness, and why most of our happiest moments involve other people. It was lovely, but it was also, for me, a new experience. People sometimes think of philosophy as an activity conducted only by well-educated deep thinkers, something beyond everyday people. This is false—and dangerous, to boot. As long as we can reflect on our experiences, we can do philosophy.

Is There Life after Death?

No. Literally every waking minute of your life gives you fresh evidence that your experiences, your inner life, everything that goes on in your mind depends on what happens to your body. As neuroscience proceeds along at a steady clip, we get more detailed evidence of essentially the same conclusion; each of the processes that make up your mind would stop in its tracks but for the full functioning of some routine in your nervous system. As philosophers like to say, we know that the mental supervenes on the physical: nothing changes in your mind without something changing in your body.* So when your body is destroyed, there won't be

---

* Quick digression: this isn't quite right. Suppose that I have middle-stage dementia and I am very competent around my own house: I can cook, do laundry, and so on. Recognizing this, my family members work hard to keep me at home. But if I were to move elsewhere, I would lose all these skills instantly. Here's the situation: I remember various things if I am in a cooperative environment, and I wouldn't remember those things if I were in an uncooperative environment, even though my body wouldn't change between those two environments. In other words, my mental states supervene, not just on my body, but on my physical surroundings. In this way, my mind might be "extended" into my environment, as philosophers sometimes say.

I find this idea incredibly interesting, but it doesn't really affect the point I'm making about life after death."

anything for your mind to supervene on.* So you won't have a mind to speak of.

Of course, people say things like: "You'll live on in the memories of others" and "Your energy will linger in the world forever." These things can be comforting, but they're not very convincing substitutes for immortality.

This is sad. But there's an interesting and hopeful nearby phenomenon. When people reflect on their mortality, they often change their priorities in sometimes surprising ways. The rapper Nas tells us: "Life's a bitch and then you die. That's why we get high / Cause you never know when you're gonna go." Philosopher-psychologist Karl Jaspers tells us that we should focus on experiences of eternal things. Medieval Christians tell us (what else?) to be better medieval Christians. Shakespeare, in "Sonnet 73," tells us to be better lovers. Ashley Madison, the dating service for people already in relationships, tells us: "Life is short. Have an affair."

These characters draw very different practical conclusions from their own mortality. But I think most of them have something in common. They seem to be aiming for the next best thing; that is, they seem to be aiming for something that is good in a way that is similar to the way that living forever (or at least for much longer than we do in fact live) would be good. I don't know whether any of these people are right or wrong but reflecting on our own mortality clearly throws some of our own most cherished values into relief. As the philosopher Steven Luper puts it, "Disquiet concerning death is the other face of love for life."

---

* Okay, just one more digression: suppose that I get an ache in my thigh and I think, "Oh, no! I have arthritis of the thigh." I go to the doctor, who informs me that arthritis is a joint condition, not something you can get in your thigh. But now consider a nearby possible world, where my whole life story is exactly the same, except the medical community uses "arthritis" to refer not only to joint inflammation, but also to inflammation in the tendons. Again, I think, "I have arthritis of the thigh." But this time I'm right! That is, my whole life story, the sequence of internal states of my body, all of that hasn't changed, but the meaning of my concept of arthritis has changed. In other words, my thoughts supervene, not just on my body, but on the communities of experts whose usage determines what those thoughts mean. Philosophers sometimes call this "semantic deference."

Again, super interesting, but shouldn't make anyone more optimistic about the prospects for life after death."

Are Science and Religion Compatible?

There are really two questions here, although they can be hard to tease apart in any given case. The logical question is whether scientific and religious beliefs are logically consistent or contradictory or whether they render each other more or less probable. The practical question is whether religious practice, in an individual or in the wider society, fosters or thwarts the practice of science.

The answer to both questions is: all of the above. Logically, if any set of scientific beliefs is consistent and any set of religious beliefs is consistent and the two are framed in different vocabularies, the union of those two sets is also consistent.* So if, for example, my religious beliefs only contain moral and supernatural vocabulary and my scientific beliefs contain no such vocabulary and both are internally consistent, then my religious and scientific beliefs are jointly consistent. When the biologist Stephen Jay Gould talked about the "non-overlapping magisteria" of science and religion, this is the sort of thing he had in mind. I don't doubt that some people's religious beliefs are like this or at least that some people aim to make their religious beliefs like this, But we can also find all sorts of historical examples of religious beliefs that were either later overturned by

---

* Google "Craig's interpolation theorem" if you want the proof of this.

science or came into conflict with the science of their day. A quick perusal of A. D. White's *A History of the Warfare of Science with Theology in Christendom* shows that at various points in time people have offered religious reasons to believe that there are no people on the other side of the globe, that the earth (and Jerusalem in particular) is the center of the universe, that each species was created separately, that water covers only one-seventh of the surface of the planet, that the world is less than six thousand years old, and that the distribution of fossils is a result of Noah's flood.

It's slightly trickier to say when scientific beliefs and religious beliefs, while consistent, render one another more or less probable. On the one hand, when the astronomer Johannes Kepler first learned that other planets in our solar system were basically like earth, orbited by their own moons, he inferred, for broadly religious reasons, that they must be inhabited. While today religious people have reconciled themselves to the fact that all of the planets we have observed don't seem to contain anything like the variety of life found on earth, Kepler's inference suggests that some religious beliefs and belief in a largely lifeless universe are at odds. On the other hand, religious and scientific beliefs are sometimes mutually reinforcing. In the century before Newton, many scientists thought that the only legitimate scientific explanations of physical processes were in mechanical terms—that all motion was ultimately to be explained in terms of (often microscopic) gears and pulleys and screws pushing and pulling on one another. The mechanical philosophy, as it was called, came to be so popular in part because it fit so neatly with the congenial theological view that God designed the machine of the world, wound it up, and let it run. In our own time, psychologists have found support for certain Buddhist teachings about meditation and attention. (Although it's a bit contentious to describe Buddhism as a religion.) In short: science and religion can be completely logically incompatible and a perfect fit and everything in between, depending on what science and religion we're talking about.

As for whether science and religion are practically—and not just logically—compatible, there have been many individual scientists whose scientific practice is religiously motivated or informed. In the work of early modern physicians and "natural magicians" like Paracelsus, the empirical and the occult are hard to disentangle. Theistic scientists have often thought of their research as natural theology or an investigation into the mind of God through the natural order of things. But we can also find examples of people whose religion has discouraged them from practicing science, from Blaise Pascal, who more or less gave up a remarkable scientific and mathematical career after his religious conversion, to all of the people who might have been scientists if they hadn't been exposed to antiscientific religious propaganda. At a society-wide level, we find a similar range of cases. The sociologist Robert Merton has argued that modern experimental science arose in England when it did in part because of Protestant culture, and heliocentrism became popular in non-Catholic countries when it did in part because it allowed them to score points against the Catholic Church for its shabby treatment of Galileo. But it's not hard to think of examples of religious institutions or prevailing religious attitudes suppressing scientific investigation; stem cell research in the United States today comes to mind.

Of course, you might want to know not whether some historical science is compatible with some religion but whether the *best* science is compatible with any religion you might be interested in practicing. I'm not in a position to answer that for you, both because I don't know enough about science and because I'm not you. But now seems like as good a time as any to ask why you *want* religion, if you do.

Is There Objective Truth?

hen people ask whether there is objective truth, they can have a few things in mind. Sometimes they're really asking not about truth but about our beliefs—whether anything is, as a matter of fact, universally agreed upon or whether our beliefs are all somehow contaminated by our social positions and life stories. Sometimes they really are asking about truth, though. It might be whether anything is true full stop or whether truth is always relative to a person, a culture, a conceptual scheme, or something like that. It might be whether there are mind-independent truths—whether what makes a belief or statement true is always, in part, how people think about or treat its subject matter. Let's take each of these questions one by one.

I don't know whether anything is universally agreed upon, but I'm not sure why it matters. If somebody out there thinks that one and one is three, does that make any difference to what I should think or do? (There is a more serious nearby problem, though, which philosophers sometimes call the problem of genealogical anxiety. If you realize that, say, you're only a liberal or you only believe that the sun is at the center of the solar system because you live in the time and place that you live and not from your own independent assessment of the evidence, that might reasonably shake your confidence in these beliefs. But the problem isn't that *some* people out there disagree with you; it's that you believe much of what you believe for reasons that have nothing to do with the truth of those beliefs.)

Whether all of my beliefs are contaminated by my particular life conditions is another fuzzy question. Here are a few types of belief contamination someone might have in mind:

+ The person has a stake in whether the belief turns out to be true or false.

+ The belief was arrived at emotionally.

+ They were persuaded of the belief by evidence that other reasonable people wouldn't accept.

+ It's not really a belief but an emotion or some other sort of attitude.

If you think about it, you can come up with examples of beliefs that are more or less (if not completely) uncontaminated—or objective—in each of these senses. But again I think there's a better question right around the corner. Why *should* we value any of these types of objectivity and under what conditions? For example, sometimes you want to consult an impartial judge, someone who has no skin in the game. But sometimes people who are personally invested in some matter understand it best. (I'm thinking of prediction markets.) Why sometimes the one and sometimes the other?

As for whether anything is true full stop or whether it's always relative to something, it's helpful to think about things that are more or less uncontroversially relative and why we think they're relative. Left and right and the time of day are uncontroversially relative (to a perspective and to a time zone, respectively). One way you know that these things are relative is that this explains some facts about disagreement. When I say (in Brooklyn) that it's 1:00 p.m. and my brother (in California) says that it's not 1:00 p.m. but

10:00 a.m., someone who didn't know about how people tell time might think that we disagree. But contrary to appearances, we agree completely. Relativity is what explains this agreement. Another, trickier type of relativity can be found in Einstein's special theory of relativity. It's a consequence of the theory that simultaneity is relative; two events can happen at the same time relative to one frame of reference and can happen at different times relative to another. The way we know that simultaneity is relative is that this is a consequence of the theory, and we have lots and lots of evidence that the theory is true.[*] That is, the theory makes all sorts of novel, surprising predictions about how various observations and experiments will turn out, and those predictions have (within some limits) all turned out right.

So we have two ways of figuring out whether something is relative: either from some facts about disagreement or from empirical support for a relativistic theory. We don't have either of those reasons for thinking that truth is relative. If one person says that such and such is true and another person says that such and such is not, they typically disagree. And no theory that I know of makes a bunch of successful, novel, surprising empirical predictions and also entails that truth is relative. So as far as I know, we have no reason to think that truth is relative.

---

[*]   I'm skipping over a fun philosophy of science problem here, which is, coincidentally, a nice way into thinking about the compatibility of the scientific picture of the world and common sense. Before Einstein, most people thought that simultaneity was absolute, not relative. But what exactly did Einstein discover about simultaneity? Einstein says that we discovered that simultaneity is relative to a frame of reference. But we could have said that Einstein discovered that there is no such thing as simultaneity—that there is only relativistic simultaneity. We could also have said that Einstein discovered this other phenomenon, relativistic simultaneity, but that two things can still be absolutely simultaneous—namely, if they're more or less relatively simultaneous for a certain set of frames of reference. Why do we interpret his theory the way we do? (We could ask this same question about any other scientific discovery that leads us to revise or reinterpret our commonsense categories.) For the record, I have no idea.

What about the idea that the truth is mind dependent, that whether something is true always depends on how we think and act? There are certainly *some* mind-dependent truths. The most obvious examples are truths about what we think and do—that I enjoy philosophy, that I took my dog for a walk this morning, and so on. Some more interesting examples involve what we might call social or artifactual kinds—whether something is a chair or a dollar bill or a paperweight or a library depends on how people think about and interact with that thing. Maybe most of what we worry about most of the time is mind dependent. Maybe some things that aren't obviously mind dependent really are—facts about morality or gender or race or color or the direction of time, say. But the question is whether *every* truth is like that.

I think the answer is no. If a truth is mind dependent, it wouldn't be the case if there were no minds or if we thought and acted completely differently than we actually do. So if it were a mind-dependent truth that, say, water freezes at 0° Celsius (at 1 atm of pressure), then if people thought or acted differently or if there were no people around at all, water would freeze at a different temperature. Of course, the word "water" might have meant something different (or nothing at all) in English, or we might have used different units of measurements for temperature. But that's beside the point. Water froze at 0° Celsius before "water" meant *water*, before we used Celsius to measure temperature, and before human beings came into existence. (If you don't believe me, check the ice sheets.) So there is at least one mind-independent truth. I'm sure you can come up with more examples on your own.

Long story short: if any of these things is what you mean by "objective truth," we don't have to (What, me?) worry whether there are objective truths. But we're not quite out of the woods. The questions people frame in terms of objective truth are *this close* to some other philosophical problems that are still wide open—genealogical anxiety, why and when we should value the several flavors of objectivity in belief, what it takes to convince us that something is relative, the problem of just how much of our everyday concerns are mind dependent. If you find yourself in a conversation about objective truth, I recommend politely but firmly changing the topic to any of these more pressing questions.

What Is Happiness?

A little ground clearing: different people mean different things by "happiness," but I'll treat this as a question about *subjective well-being*—the aggregate of the psychological features of a person that (perhaps partly) constitute how well their life is going for them. Note that if how well your life is going is partly a matter of what's going on outside of your mind (say, if you have the respect of your friends and colleagues), that's not happiness as I'm using the word. Note also that I take happiness to be whatever is going on in your mind that your well-being (partly) *consists in or reduces to*, not whatever is going on in your mind that causes or is correlated with your well-being. So it could be that listing the things you're grateful for causes you to be happy or that people who have a certain number of close friends are happy. But those things don't make for a happier life all on their own. Note, finally, that I'm thinking of a happy person as someone who has the psychological components of a life that's going well *for them*. Godzilla might be happy stomping through the streets of Tokyo, even if what makes him happy doesn't work out well for the locals.

But that just makes the question a little more precise; it doesn't answer it. To get there, it's helpful to distinguish four theories:

The Hedonic Theory: Happiness is pleasure and the absence of pain.

The Preference Theory: Happiness is getting what you want.

The Life Satisfaction Theory: Happiness is being satisfied with your life.

The Emotional State Theory: Happiness is a set of emotions and emotional dispositions.

I don't have a strong view here, but I lean toward the hedonic theory. First, it seems like bedrock that if my life is more pleasant than yours, other things being equal, mine is going better for me. Second, it stands up pretty well to some compelling objections to the other theories.

One problem with the preference theory is that people want all sorts of things that intuitively don't make their lives better. This could be because they're mistaken about the likely effects those things will have on them. ("I thought I would feel better if I drank my fifth cup of coffee for the day, but I turned out to be wrong.") It could also be because our desires can be manipulated by features of our environments that intuitively have nothing to do with how well our lives are going for us—keeping up with the Joneses, wanting the newest gadget when the current one is just fine, other so-called endogenous or adaptive preferences.* And even if your preferences don't rest on any kind of mistake, some of them are, in a sense, entirely on behalf of other people. I'd like to see the literacy rate in Canada improve, but only because it would be good for the people of Canada. It wouldn't affect me much at all.

These problems don't really apply to the hedonic theory. If I'm mistaken about the likely effects of some pleasant experience, I'm still better off, other things being equal, for having the pleasant experience. If I've been trained or manipulated not just to want something but

---

* Some philosophers have tried to defend the basic idea of the preference theory by counterfactually idealizing away from our ignorance and irrationality. Perhaps happiness isn't getting what you want but, say, getting what you would want your actual (more or less ignorant and irrational) self to want if you were perfectly rational and maximally well informed. This is a pretty weird counterfactual, though; there might not be a fact of the matter about what this supersized version of yourself would want you to want. And this also doesn't solve the next problem with the preference theory.

to take pleasure in it—say, by my early childhood exposure to spicy food—I'm still better off, other things being equal, for having that pleasure in my life. And I can pick out the preferences that have something to do with *my* well-being by looking for preferences it would be pleasurable to satisfy.

There are a few problems with the life satisfaction theory, too. One person could be more satisfied with their life than another not because things are going better for them but because they have really low expectations. (If someone faces nothing but boredom, pain, and humiliation day after day, but they think this is what they deserve or the most they could hope to get out of life, they might be very satisfied with their lives, but they won't be happy.) It's also really hard to keep track of everything that happens to you, and so how satisfied you are with your life at any given moment could just be a matter of the first satisfying experience that comes to mind.

Again, no problem for hedonism. Someone can't, as a rule, instantaneously make their life more or less pleasurable simply by changing their expectations. And to the extent that your expectations *do* affect how much pleasure and pain you experience (say, if you set yourself up for disappointment by being overly confident that you will get some job you've applied for), it's not implausible that you could be happier than someone with an otherwise similar life just by expecting things to turn out differently. Again, being satisfied with your life requires you to make a really complicated all-things-considered judgment about how things are going, which could be distorted in various ways. How much pleasure or pain you've experienced isn't, in this way, a matter of judgment; it's a matter of how your life is *really* going, independent of what you may think.

Lastly, there's a lot to say for the emotional state theory. It doesn't make your happiness implausibly dependent on your expectations, on some tricky evaluation of your entire life, on your possibly misguided preferences, or on what you want for other people. And it's hard to doubt that a life with more positive emotions is going better than a life without, other things being equal. But the problem with the emotional state theory is that some things that make your life better aren't emotions. Stubbing your toe might make you angry, but the pain in your toe isn't an emotion. Having an orgasm might endear your partner to you, but the orgasm itself isn't that feeling of warmth or love or anything like that. Perhaps these things make less of a difference to your well-being than emotions, but they still make some difference.

One sort of case that emotional state theorists use to argue for their view is a person who is deeply depressed but manages for some period of time to distract themselves from their depression with a series of fun or engaging activities. They have a pleasurable life, but it seems weird to call them happy. I don't disagree, at least for *some* sense of "happy." But I think we have a tendency to describe anything that's regrettable or inadvisable about someone's life as "unhappiness." And there's clearly something regrettable about the distracted depressive. They're not as happy as they would be in the long run if they dealt with their depression more directly. And they're in an extremely *fragile* state; the slightest interruption in their series of distractions, and the depression is going to hit them like a ton of bricks. Their luck is going to run out eventually. Those things are all consistent with saying that yes, for now, they have high subjective well-being; they're happy, in the sense I'm interested in.

The main problem for hedonism, at least that I'm aware of, is that there doesn't seem to be a single feeling of pleasure or pain. A peaceful walk in nature, the feeling of being well rested, a good workout, learning something interesting, eating a delicious meal, and hearing a funny joke are all pleasant experiences, but do they really have anything in common? I'm not sure how serious this problem is, though. It's certainly hard to describe *at length* what all pleasurable experiences have in common, but they do have one pretty clear thing in common—namely, that they're all pleasurable experiences. I mean, maybe pleasure and pain are extremely heterogeneous; maybe different types of pleasure are incommensurable within certain limits. (What's the going exchange rate between clever puns and solid naps?) But so it goes—then different types of happiness are, too.

We might also object to hedonism on the basis that we really *care* about happiness, and pleasure seems small and insignificant by comparison, so happiness must be something more than pleasure. But this objection only seems plausible if you have small and insignificant pleasures in mind. It's pleasurable to fall deeply in love, to accomplish something meaningful with great difficulty, to have an intimate, life-changing conversation. To call these things pleasures isn't to sell them short; it's just to remind us that pleasures aren't all little things.

I was talking with one visitor to the booth about the experience machine. Suppose you could go in a machine that could give you the experience of anything you wanted. If you want to be a famous violinist, for example, it would give you the life of a famous violinist. These experiences wouldn't be real, though, and you have to stay in the machine for the rest of your life. Setting aside your responsibilities in the real world, would you go in the machine? (And if not, does that mean that there's more to well-being than happiness?)

The person I was talking with had a great argument for no. The way we find out what makes us happy is by getting thrown into new situations, trying new things, and seeing how we like them. We have to discover what we like. When you go into the experience machine, you only bring the likes that you've already discovered. If the experience machine only gives you these things, wouldn't you risk missing out on all of the joys you haven't learned of yet?

Is There an Unconscious?

If the question is whether some of our mental states are unconscious, the answer is yes. A mental state is unconscious, roughly, if you don't know about it just by attending to it. Whatever you know about it, you learned indirectly, by inference. In this sense, perhaps most of your mental states are unconscious. For example, if someone asks you to figure out whether something is a bird, you will probably respond more quickly and accurately if they show you a picture of a pigeon than if they show you a picture of an ostrich. You have a stereotype of a bird, and pigeons fit that stereotype better than ostriches. This stereotype is a sort of mental state. But you can only learn about it indirectly, by (among other things) trying to categorize things as birds and observing what happens. So this stereotype is unconscious.

More dramatically, some people with a certain sort of brain lesion are, as far as they can tell, blind in one half of their visual field. If you place an object in this half of their visual field and ask them what it is, "blindsight" patients will, very reasonably, tell you that they don't know. But if you make them guess, incredibly, they answer accurately. And if you put an obstacle on their blind side, they'll walk around it. So they can sense the objects on their blind side, but they can only find this out by observing their own behavior. (Blindsight raises a whole bunch of other interesting philosophical questions. When a blindsight patient finds out that there's a pencil in front of them but only in this sort of indirect way, do they know that there's a pencil in front of them? What does this tell you about knowledge? Blindsight shows that you can make certain uses of visual information when it's unconscious but not others. What does this suggest about what consciousness is for?)

Today, if people doubt the existence of the unconscious, my bet is that it's because they doubt a particular theory of what goes on there—Freud, your mom, potty training, the whole bit. But in the very recent past, philosophers and psychologists thought that all mental states were conscious pretty much by definition. Even though most philosophers and psychologists don't think of the unconscious the way Freud did, we probably have him to thank for the fact that the unconscious is now just a piece of common sense. But like a lot of common sense, it both isn't very common at all and also has this weird feeling of inevitability. The unconscious has the good fortune of being backed up by the evidence, but it's probably a healthy habit to be on the lookout for features of common sense that don't enjoy the same scientific support.

Some families with a gaggle of elementary school-aged children stopped by the booth, and one of the moms asked her daughter if she had any philosophical questions. "Who's better, Freud or Jung?" It turned out, in the ensuing conversation, that this kid knew a remarkable amount about the two philosopher-psychologists and had plenty of interesting things to say for herself about how well we know our own dreams and what dreams do for us. Still, I couldn't help but find it a little sad that somehow, even at that age, she had already gotten the sense that asking about philosophy meant asking about some great philosophers. I mean, there's a place for talking about the Big Brain Boys of the past and for trying to understand other people's thinking, more generally. But please, please, please: take your own thoughts seriously.

How Can We
Overcome Dualism?

W hen I first started doing the Ask a Philosopher booth,
I think I expected most people to come with questions
about social and political philosophy and their personal
ethical problems. After all, these are the philosophical problems that
people *have* to face just by reading the news and trying to get along
with the people around them. But I was wrong. When you give people a chance to talk about what's on their mind, you find that even
the most abstract or speculative or otherwise impractical questions
come up again and again. Somehow, they get in the air. It's a hopeful but pointed image—all of these people walking around the city
wracked by their seemingly weird, private problems, unaware that
the people around them are thinking about more or less the same
thing. Anyway, two people asked this question, in so many words, at
consecutive booths one week apart.

Obviously, a lot depends on what we mean by "dualism." If dualism is the tendency to think in black and white or in exclusive and ex-
haustive categories, the problem with dualism is clear enough: it makes
subtlety invisible. One way to fight back against this kind of dualism is
to bear this problem in mind and pay attention to gradations and con-
tinuities and subtleties when you can. Perhaps sometimes you should
reject the scale or continuum altogether. Maybe something isn't good
or bad, healthy or sick, mental or physical, natural or artificial, feminine
or masculine, and maybe it isn't anything in between those.

But I don't think this is what the visitors to the booth were worried about. Dualism, for them, is the tendency to think of things as separate that are deeply, intimately bound up with one another. This might take on a theological cast—thinking of God as transcending the world rather than in the world. It might also reflect a desire to sustain the oceanic feeling of mystical experience, the sense that the differences between things or between you and the world around you are somehow conventional or illusory or otherwise less meaningful than they appear.

There are a few ways of overcoming dualism in this sense. The Eleatic philosophers of ancient Greece tried to argue, on the basis of pure reasoning, that the apparent differences between things were all illusions—that there is literally only one eternal, unchanging thing. Zeno's paradoxes of motion are the most famous contributions to this genre. In order for an object to move from one point to another, first it has to travel half the distance, then half the remaining distance, then half the remaining distance, and so on. Since it will have to travel an infinite number of distances in order to reach its destination, and that's impossible, motion is impossible.

But these sorts of arguments are unlikely to convince anyone who isn't already convinced of their conclusion. (For starters, there's a difference between traveling an infinite distance and traveling a sum of an infinite number of distances.) I don't doubt that there are pieces of common sense that are logically contradictory.* But there is nothing logically inconsistent, at the bare minimum, in supposing that there is more than one thing in the world.

Another approach is to argue that although things are distinct from one another, there is one thing that everything else is a part of or grounded in. For example, suppose for any two objects there is a whole that has those two things as parts. Then you might think of the universe itself as the whole that has every other thing as a

---

* Absence makes the heart grow fonder, but out of sight, out of mind?

part. So even things that seem completely unrelated have at least that much in common: they are both parts of the same whole.

Something like this thought is elaborated at greater length and in a way that might better articulate the metaphysical insight people seem to get from their mystical experiences, by the seventeenth-century philosopher Baruch Spinoza in his *Ethics*. (At least if I understand him correctly.) Many philosophers are interested in the fundamental structure of the world: what are the most basic building blocks of reality and how are the objects that we're familiar with built out of those blocks? Well, what makes a building block "basic" in the relevant sense? One thought is that it is what philosophers used to (and I guess sometimes still do) call a *substance*—something that is capable of existing independently of anything else. Spinoza argues that there is exactly one substance (God or Nature, whatever you want to call it).* This substance has some attributes, among which are Thought and Extension (or Mind and Space). These attributes themselves have various "modes," and the modes are the facts of everyday life. What is it for me to be tall, in Spinoza's theory? Roughly: God extends in an Ian-is-tall kind of way. In other words, the everyday world that we experience consists of an assortment of properties of properties of God, and God is the only fundamental thing. I don't know what to say about the merits or otherwise of Spinoza's theory, but I think it does a good job of making room for the thought that nothing is really separate from anything else, without requiring you to give up—in an implausible or humanly impossible way—your basic beliefs about how the world works.

Lastly, we can try to remember the more or less contingent, empirical connections that tie various sorts of things together. It's a familiar point—but only because it's so affecting—that everything on earth

---

* Spinoza's argument is an intricate, fragile piece of reasoning. But there's at least one blunt, somewhat intuitive reason to think this might be true. If anything exists, the totality of all things exists. So nothing can exist independently of the totality of all things. So there is at most one substance—namely, the totality of all things.

is made of stars. Every species depends for its continued existence on countless others. One common ancestor of all placental mammals— you, your dog, the blue whale—lived, by some estimates, just 65 million years ago. Every human structure rests on wilderness and is at most a few years of neglect away from being completely reclaimed by that wilderness. Almost everything you know you learned from the testimony of other people. With a handful of possible exceptions, every human adult alive today has been the beneficiary of thousands and thousands of hours of childcare. And so on. None of this is news to you, I'd bet. But then why is all of it so easy for us to forget?

One of the people who brought up the dualism question offered a gorgeous interpretation of the Adam and Eve story. We usually think of this as a story about human frailty or ingratitude or disobedience: Eve just wasn't strong or good enough not to eat the apple, and she brought Adam down with her. But, the visitor suggested, this doesn't really address the fact that she ate from the tree of the knowledge of good and evil. Maybe Eve's mistake wasn't that she succumbed to the temptation of the snake or that she forgot what God had told her but that she came to believe in good and evil at all. Maybe this little dualism is the source of all of our troubles.

I don't actually believe this. (Something about the best lacking all conviction.) But I understand the story better than I did before.

Are Space and Time
Objectively Real?

H oo boy.

Let's treat this as a question about whether space and time are, as they say, absolute or relational. The stock example of an absolute theory of space is Newton's. He thought that there is a single frame of reference or coordinate system where the velocity of an object in that frame of reference is the object's *real* velocity. Any two points in this reference frame are always the same distance apart. For Newton, the most fundamental facts about motion are facts about how objects change location over time in this frame. While the space that is described by this coordinate system isn't exactly material, it is a thing of some sort, which is the way it is independently of what goes on inside of it. It is at least conceivable that things could move relative to this coordinate system even if they don't move relative to any other objects—say, if the entire universe were moving at a steady one kilometer an hour in a certain direction.

One of Newton's reasons for believing in absolute space came from his famous bucket experiment, which you can try at home. (You can also see it illustrated on the next page.) Fill a bucket halfway with water and suspend it from a string. Turn the bucket around a bunch of times so that the string is twisted up tightly. Steady the bucket, then let it go. At first, the bucket will rotate faster than the water, and the surface of the water will remain flat. But after a little while, the water will start rotating as fast as the bucket, and the surface of the water will become concave. In other words, the surface of the water is flat when the bucket is rotating with respect to the water, and it is concave when the bucket is *not* rotating with respect to the water. So the surface of the water isn't changed when the water is moving relative to the bucket, and it is changed when the water is *not* moving relative to the bucket. Newton inferred that the type of motion that causes the water to become concave can't be relative, so it must be absolute; that is, it must be rotating in absolute space.

Newton's Bucket Experiment[*]

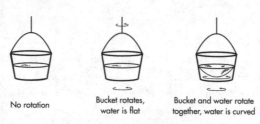

No rotation

Bucket rotates, water is flat

Bucket and water rotate together, water is curved

We can approach the question whether space is objectively real by asking whether Newton's theory was right. The theory makes a few different but related claims, which we can tease apart:

Singularity: There is a frame of reference that is more fundamental than all the others for the purposes of mechanical explanation.

Eternity: There is a reference frame in which the distance between any two spatial points is the same across all times.

Priority: Facts about the spatial relations between objects are, in principle, ultimately explained in terms of facts about the locations and velocities of those objects, not vice versa.

---

[*] Adapted from Jeroen van Engelshoven, "Study on Inertia as a Gravity Induced Property of Mass, in an Infinite Hubble Expanding Universe," *Advances in Mathematical Physics* (2013).

Substantivalism: Space exists independently of what goes on inside of it.

Which of these claims hold up?

Singularity is false because all inertial frames of reference are on a par, explanatorily. The most parsimonious modern explanations of the bucket experiment don't appeal to absolute space. (Roughly, because when the water becomes concave, it is rotating in any inertial frame.) And other attempts to identify a single privileged frame of reference have failed for other reasons. (For example, Hendrik Lorentz developed a theory that makes all the same predictions as Einstein's special theory of relativity but privileges the frame of reference in which the "luminiferous ether" is at rest. The problem is that there is, to all appearances, no such thing as the luminiferous ether.)

Eternity is false because in the general theory of relativity the distance between two spatial points varies with the curvature of space, which changes whenever the distribution of matter in the universe changes. Since the distribution of matter in the universe is constantly changing, the distance between any two points in space is constantly changing.

The jury's out on priority. There are some "relationist" ways of stating the theory of relativity that put spatial relations between objects first and define the other relevant concepts in those terms. But there are other experimentally indistinguishable ways of stating the theory that define things the other way around. (It's a bit like the difference between defining a sister as a female sibling and defining a sibling as a brother or sister.) The considerations that pull you in one direction or the other are sort of mushy. At least there's no consensus about which approach is correct, and I definitely don't have an opinion one way or the other.

The jury's also out on substantivalism. If space and time are dimensions of a single four-dimensional space-time, they don't exist independently of one another. And if the curvature of space-time depends on the distribution of matter in space, the specific geometry of space-time depends on what happens inside of it. But there are some properties that space-time might have independently of what events actually take place—for example, if there is a nonzero cosmological constant, so that even an empty universe would tend to expand. In any case, there's no agreed upon interpretation of substantivalism, but *if* we take it to be the view that space-time has this or that property regardless of the events that actually unfold within it, it becomes something we can, in principle, investigate empirically.

OK, glad that one's over. I think we exhausted all the physics I know about three paragraphs ago.

Why?

When people ask this sort of thing sincerely, it's a bit like turning on a white noise machine. We have some sort of intellectual discomfort, and we hope that someone can get us out of it or at least distract us from it by talking. There are more precise questions out there, but I'm happy to oblige.

There are a few flavors of answer the questioner could be in the market for. We could ask about what, if anything, *justifies* any of our daily actions. (Why get up in the morning?) We could ask about the *purpose* or *meaning* of this or that. (Why are we alive?) But you've probably already thought about those things a fair amount. So let's try a less well-known and potentially more exciting way of interpreting the question: what makes for an explanation, in general? That is, when a why question asks for an explanation—and not a justification or rationalization—what counts as a correct answer?*

Explanations can give us understanding, guide us in making novel inferences, and (at least when we construct them ourselves) help us retain new information and learn new skills. A large part of the scientific enterprise is devoted to crafting and testing explanations of

---

* A wrinkle: explanation-seeking why questions come in different flavors. In particular, explanation-seeking why questions are often (always?) contrastive. To take an example from the philosopher Peter Lipton, it's one thing to explain why the leaves turn yellow in November (rather than in January); it's another thing to explain why the leaves turn yellow in November (rather than turning blue). What counts as an answer to a why question depends in subtle, often tacit ways on what, exactly, we're trying to explain.

things. So we clearly *care* about explanation. But it's pretty hard to say what it is to explain something, as opposed to merely describing or predicting it. After all, there are lots of different explanations out there, and it's not clear what they have in common.

One way to get a better understanding of how explanation works is to consider one really influential theory of explanation, what it gets right, and what it gets wrong. In the twentieth century, the philosophical discussion of the nature of explanation mostly traces back to the deductive-nomological theory (DN).* According to DN, an adequate explanation of some fact is a deduction of that fact from some laws of nature and, perhaps, some additional facts about how they apply in some circumstances. For example, suppose I'm playing a black-and-white version of Tetris and I have to figure out whether one block is a reflection or a rotation of another. It takes me longer when the block is rotated 180° than when it's rotated 90°. Why? There's a law of nature at work: when people have to figure out whether one shape is a rotation of another, the amount of time it takes them to answer is an increasing linear function of the number of degrees that the shape has been rotated. DN also offers a good account of how we explain not just specific facts but laws of nature themselves. For example, the general theory of relativity can explain why Newton's laws of gravity and motion apply pretty accurately in our solar system because Newton's laws are a so-called limiting case of Einstein's; they obtain, more or less, when nothing is moving very quickly and space-time is locally pretty flat or when you assume that the speed of light approaches infinity. DN also explains why there can be more than one explanation of something—namely, that there is more than one way to deduce any given fact from some laws of nature.

---

* A deductive argument is a chain of reasoning where, unless the reasoner has made a mistake, the premises entail the conclusion; if the premises are true, the conclusion has to be true, by virtue of the structure of the argument alone. "Nomological" means having to do with laws.

Zooming out a bit further, DN makes sense of how the relatively murky business of explanation fits into some other, tidier scientific enterprises, like describing and predicting. An explanation is a sort of prediction based on a law of nature. Insofar as laws of nature are stable features of the world, what we can explain is, in a sense, what we should have predicted all along. If you, like me, are uncomfortable with mysteries, this is a nice feature of the theory.

So DN seems to fit *some* types of explanation pretty well, and it has some more abstract theoretical virtues. But it has some important limits:

IDEALIZING EXPLANATIONS

Some widely accepted explanations start from premises that we know are literally false. For example, we might use Boyle's law to explain why pressure increases as the container of some gas shrinks. But Boyle's law only applies exactly to "ideal gases," which don't really exist. Or biologists might explain how some trait spreads through a population of rabbits using a model that assumes that the population is infinitely large. Maybe this isn't a serious problem for DN, though; we could just add a premise to the effect that this gas behaves more or less like an ideal gas or that the model more or less accurately captures such-and-such features of our rabbits. That does raise another issue, though.

LAWS OF NATURE

What exactly is a law of nature? One standard answer is that laws of nature differ from other universal generalizations in that they support counterfactuals. For example, it's a law that the tide in a certain region of the earth will be highest when that region is either closest to or opposite the moon. So this also holds counterfactually: if some region that is not currently opposite the moon *were*, then it would be high tide there. (Compare: all of my immediate family members are more than four feet tall. But this doesn't hold counterfactually; if a baby were a member of my immediate family, they would *not* be more than four feet tall. So it's not a law of nature.) But if this is right, it's pretty weird for a couple of reasons. First, counterfactuals are a sort of mushy, vernacular matter. If they play an important role in defining what counts as a scientific explanation, then what appear to be our least mushy, most hard-nosed, mathematically sophisticated scientific practices would wind up pretty mushy after all. Second, counterfactuals are, on their face, statements about what goes on in *other* possible worlds. Why should facts about the laws of nature in our own world—or what explains what—depend on what goes on elsewhere? Maybe there's a way around this, but we'd need another definition of "law of nature" to find it.

EXPLANATORY ASYMMETRY

Take the moon and the tides again. If I want to know why it's high tide somewhere, I can give a perfectly good explanation by citing the law I just mentioned and adding that that place is currently opposite the moon. But what if I want to know why the moon is opposite that place? Well, I could deduce it from the law in question, along with the fact that it's high tide there (and it's not as close as it will be to the moon on that day). But that's silly. The position of the moon explains the tides; the tides don't explain the position of the moon. That seems to have something to do with the direction of causation: gravitational waves go from the moon to the water on the surface of the earth, and that's what causes the tides to rise and fall. But nothing the tides do makes much of a difference to the moon.

EXPLANATORY RELEVANCE

Let's suppose that laws of nature are generalizations that support counterfactuals. To take a classic example, suppose I'm biologically male, but I want to make extra sure that I never get pregnant, so I take (female) birth control. This is in accordance with a certain law of nature, namely, that no biological males that take birth control will get pregnant. So I have an explanation of why I never get pregnant:

no biological males that take birth control will get pregnant, and I'm a biological male who takes birth control. But of course the fact that I take birth control is irrelevant. I wasn't going to get pregnant even if I stayed off the pill. So a correct explanation of some fact has something to do with identifying what is, so to speak, explanatorily relevant to it. Laws of nature are not explanatorily relevant to all of the facts they can be used to deduce.

NON-NOMOLOGICAL EXPLANATION

Sometimes we explain things without reference to laws at all. Why did the vase shatter? Because I dropped it—no law necessary. And even if I wanted to bring in a law here, it's not clear what it would be. After all, not every vase that is dropped shatters; perhaps most dropped vases don't shatter. Or, to take another example, we often explain historical events through narratives of a certain sort: first the Germans lost World War I, then they went through a depression, then the Nazis gained power (to simplify things just a touch). But it's hard to think what historical laws might be invoked here. Do countries that lose wars always go through subsequent economic depressions? Does economic depression always lead to authoritarianism? I doubt it. It's just that, in these particular circumstances, that's how things shook out, causally.

GOOFY PROOFS

Explanations of mathematical (and, if there are any, moral) facts raise a bunch of interesting questions. Are all mathematical theorems—or at least all theorems that are universal generalizations—laws of nature? If so, there are all sorts of trivial or unenlightening deductions of mathematical facts from laws of nature. Why is it the case that $1 + 1 = 2$? It's hardly an answer to say because it's a law of nature that $1 + 1 = 2 + 0$, and it follows that $1 + 1 = 2$. But even if we only take, say, certain axioms as real mathematical laws, not all deductions from those laws are explanatory. For example, computers that are programmed to prove new theorems sometimes cook up so-called phonebook proofs—proofs that are too long and tedious for human mathematicians to comprehend. Even if these proofs actually work, I don't know of anyone who thinks they explain their conclusions. When people talk about mathematical explanation, at least, there's a close connection to *understanding* the fact in question, which is some sort of psychological state. But merely deducing something from a law of nature doesn't guarantee understanding—or any other psychological state, for that matter.

⁜

Maybe one day we'll find a single unified theory of explanation that can deal with all these problems; maybe there are a few different types of explanation and the best we can do is understand each type. Either way, a satisfactory theory of explanation will have to account for explanatory idealization, relevance, asymmetry, non-nomological and narrative explanation, and the connection between explanation and understanding. You had the good sense to pick up this book, so I'm sure you're the person for the job.

> A little boy asked us why apples are red, and I said the first thing that came to mind, something a biologist told me once about helper pigments in red leaf lettuce. But then he put on a cheeky smile and asked his real question.
> "Why?"
> "Why what?"
> "Why?"
> His mom, who had clearly been through this a few times at home, rolled her eyes. "This guy gave you an answer to your question. You can't just say why!"
> One of the other philosophers rescued the situation by changing the subject, summarizing Aristotle's

account of the "four causes," or the four ways of answering why questions.* It was a hit.

About an hour later, a couple of guys in their twenties approached. They had been pacing around a bit, looking at the booth from a distance. Eventually, one of them came up and asked, "Why?" This time I was ready to change the subject and jumped right into the nature of explanation.

Both of them were trolling us a little. But then again, maybe a bit of friendly trolling isn't such a bad way to get a philosophical conversation started.

---

* A *material* cause is the matter from which something is made. A *formal* cause is, as far as I can tell, the form or archetype of a thing or a description of the essence of something in terms of its location in some taxonomy. An *efficient* cause is whatever preceded something in time and brought it about. A thing's *final* cause is its purpose, whatever it exists for or to do.

PART II

Personal Questions

What Is Love?

A lot of people think that love is an emotion. This is almost right. It's not an emotion, because emotions don't last as long as love tends to. You don't stop loving someone when you're asleep, busy with something else, or angry at them. But if you aren't anxious at the prospect that someone will leave you, upset when they're suffering, and at least sometimes happy to see them, then you don't love them. So love is, perhaps among other things, a disposition to feel certain things about a person in certain circumstances.

Isn't this unsatisfying, though? When people ask what love is, they're trying to find out something *deep* and *surprising* about love. If you ask me,* the idea that love is a cluster of emotional dispositions, while true, just isn't very deep or surprising. (Maybe that's why people are so shy about saying what love is.)

The trick is to find a new question. One sort of philosophical question that you can always ask when it comes to matters of the heart, which I think *does* have an interesting answer, is the question of emotional rationality. So . . .

---

* Picking up this book counts as asking me.

When Does It Make Sense for
One Person to Love Another?

We speak of emotions and emotional dispositions as apt or inapt, reasonable or unreasonable. (Think of someone angry at respectful treatment, proud of their abject failures, or fearful of what they know is harmless.) This is weird, if you think about it. When we call a belief or a statement reasonable or unreasonable, that has something to do with the evidence for the truth of that belief or statement. But love and anger aren't the sorts of things that can be true or false.

One way of thinking about when an emotion makes sense is to think about what it's *for*. Why did we evolve to have this emotion, and how has our emotional biology been fine-tuned by culture and personal development to harmonize with the sorts of lives we live today? If an emotion was prepared by biology, culture, and personal development to "go off" in certain circumstances, then it has, in a sense, misfired if it goes off in other circumstances. The question of when it makes sense to love someone, then, becomes the question of why we evolved the capacity for love and how that capacity has been shaped by where and how we grew up. If we're talking about romantic love, the evolutionary story will have something to do with pair-bonding and reproduction, and the cultural story will have something to do with the psychological and financial benefits of making a life with another person. For more of the details, we'd have to ask some biologists and anthropologists.

But there are a bunch of other interesting philosophical questions about love that we haven't touched yet: How do you know when you're in love? What, if anything, do the different types of love—romantic love, love for family members and children, love for friends, self-love, love for pets, love for activities—have in common? What would it look like to love someone authentically, uncontaminated by the expectations of the wider society? Is this even possible? Is it desirable?

Are People Born
Straight or Gay?

An old friend realized that he was gay at a young age. His parents took him to see *Aladdin* when he was around seven, and he found himself reacting differently to the hero than the people around him. He couldn't take his eyes off of him. He didn't know exactly what he wanted to do with Aladdin, but he knew that he was in love, more or less. And that was that, the story went.

My friend isn't alone of course. Plenty of gay men describe early feelings of being aroused by men and boys. (I'm sure a lot of girls were in love with Jasmine, too.) And a lot of people take these sorts of stories as conclusive evidence that their protagonists were born gay. (We don't talk quite as often about when people found out they were straight for some reason.)

While I don't doubt that these stories are basically correct and that they often play an important role in their authors' autobiographies, I think they have been slightly misinterpreted. The reason is that they are stories about people discovering that they are attracted to *someone* of the same gender. But a gay person isn't just someone who is attracted to someone of the same gender. It is (roughly and perhaps among other things) a person who more or less exclusively sexually prefers people of the same gender. The idea that people have sexual preferences for genders—rather than for types of sex acts or body types or anything else—is historically recent. The ancient Greeks, for example, thought of people as either preferring to be sexually passive or sexually active. A Greek man wouldn't have described the first time he discovered he was attracted to someone of the same gender as learning that he was gay or even that he was bi.

Of course, the mere fact that people in the past (with a few exceptions) didn't *talk* about gay people doesn't necessarily mean that there haven't always *been* gay people in roughly the same percentage as today. But if there have, that leaves us with a mystery: why did it take us so long to discover that so many of us are gay? (Remember: people have known about gay sex forever.) Here's another, less mystifying picture of sexual development: we have all sorts of innate sexual predispositions. These predispositions are honed by our experiences of sexual arousal and pleasure and channeled into the categories of sexual preference that are acknowledged in our social groups. So someone might have, say, an innate predisposition to enjoy sexual domination; as he watches porn and experiments with partners, he discovers that he enjoys certain sorts of sex acts; he learns that there's a type of person called a leather daddy, he recognizes himself in that type, and, voilà, he's a leather daddy. Perhaps some people exclusively prefer members of the same gender for purely innate reasons, but this number must be small enough that they could go largely undetected for millennia. (Not to mention that if most gay people were born gay, identical twins of gay people would mostly be gay themselves. While the evidence suggests that sexual orientation is partly heritable, this is not the case.) Most gay people probably become gay sometime after birth.

If this seems unconvincing (or offensive), that might be because of how the question has been framed in public discourse. The conventional wisdom is that you're either born gay or you choose to be gay. The idea that people choose to be gay sounds silly, because it is silly. (It's hard enough to choose whether to be attracted to a single person, let alone an entire gender.) But of course for any other trait that a person has, we would never assume that it has to be *either* chosen *or* innate. Are people born preferring dogs to cats, or do they choose to prefer dogs to cats? Neither, obviously. For one thing, we're born without any idea what dogs or cats are. For another, the thought that anyone would knowingly choose to prefer cats—and the lifetime of frustration and disappointment that entails—is absurd.

What Makes a Boy a Boy?

There are a lot of really tricky questions in this vicinity. What is the biological basis of sexual difference? How do we create and shape social identities—roles or categories that are somehow recognized by our social groups, which we use to predict and moralize about behavior? How do we create and shape personal identities—the features that are central to how we think about and express ourselves? To what extent is gender—or the various things that people have used "gender" to mean—determined by each of the above?

I don't know how to answer these questions, so I'll answer an easier question, which is I think pretty close to what the person who started this conversation had in mind: should we call trans boys boys? The answer is yes, because it is cruel not to call trans boys boys (and girls girls of course). They want you to call them boys and it is very easy to call them boys. Doing something that someone doesn't want you to do when it is very easy not to (and it doesn't make a difference to anyone else) is cruel. The cruelty is compounded by the fact that trans people are harassed and discriminated against by employers, healthcare providers, and members of the general public. Why make trans people's lives harder than they already are? Isn't this a matter of affording trans people the same basic consideration and autonomy that we grant to everyone else?

It would be nice if this were enough to convince whoever needs convincing, but I doubt it. So let's consider some likely objections.

🗣 But that's not what "boy" means.

Maybe (!) that's not what "boy" has meant. But meanings change. One good reason to change the meaning of a word is if it's harmful to a lot of people for the word to have that meaning.

🗣 But trans people are mentally ill.

What's mentally ill about identifying with a gender other than the one typically associated with your biological sex? (It's not a delusion; trans people aren't mistaken about their own biology.) But more generally, what's a mental illness? One plausible answer is that a mental illness is a harmful mental dysfunction. On this account, being trans isn't a mental illness because merely being trans doesn't harm anyone. (Of course, trans people often feel deeply uncomfortable *before* transitioning, but that's a different story.) In any case, if you want to claim that being trans is a mental illness, you need an argument based on a defensible conception of mental illness, and I haven't seen such an argument yet.

🗣 But trans people do harm others. They're dangerous in gender-segregated spaces, and they reinforce stereotyped and essentialist ways of thinking about gender.

There's no evidence that trans people harm others in gender-segregated spaces at higher rates than cis people. Something like the opposite is true, though; if a trans man who looked like me walked into a woman-only space (a locker room, say) and started undressing, that would likely upset the women in that space, with good reason.

As for the idea that trans people reinforce gender stereotypes or gender essentialism, I think the answer is: sometimes. Suppose a trans man says something like: "I've always known I was a boy. Even when I was a baby I always wanted to play with trucks and swords." This sort of claim helps perpetuate the idea that the full range of stereotypically boyish traits are naturally or essentially boyish. Of course, many trans people *don't* say things like that but granted that trans people do say this stuff sometimes. So what? Aren't the vast majority of people working daily to perpetuate gender stereotypes *not* trans? And even if trans people were more likely than cis people to do this sort of thing, what good does it do to refuse to refer to them using the nouns and pronouns they prefer?

TL;DR: the metaphysical questions surrounding sex and gender are really hard, but we could make trans people's lives a lot easier if we were decent to them.

If Someone Buys Me
a Drink at the Bar, Do I Owe
Them Anything?

A lot of our moral thinking revolves around an accounting metaphor: we "owe" people things, incarcerated people are "paying off their debt to society," and so on. The idea is that if you harm someone, they can exact a payment from you, either in the form of a punishment or in the form of a benefit. (They can also forgive you, the same way that a lender can forgive a debtor.) Conversely, if someone does something nice for you, you owe them payment, in the form of a benefit. We have to keep our moral ledgers balanced.

If someone buys you a drink at the bar and you feel the need to have a chat with them (or even to accept the drink in the first place), it looks like the accounting metaphor at work. But we should be suspicious of the accounting metaphor, both because it asks too little and because it asks too much.

It asks too little because it's deeply unambitious to think that you're in the clear morally as long as you keep the harm that you do appropriately balanced with benefit. Is this all you want from yourself or others? The world would be better off if we held ourselves to a higher standard.

It asks too much because if we take the accounting metaphor seriously, it's weirdly easy for people to impose special moral obligations on you that might be really annoying or just pointless to carry out. If Warren Buffett gave me a million dollars out of the kindness of his heart, I would be really grateful, but aren't there better things I could do with my time and money than trying to repay him? Or more to the point, do we really want to say that any lonely or horny rando at the bar can oblige you to talk to him if he has a few bucks to spare? (It sounds a little like the old con where someone bumps into you on the street with an empty bottle of liquor and asks you to pay them back for it when it breaks.)

Of course, it might be *nice* of you to talk to this person. (At least as long as you're not just encouraging him to pull the same move at the next bar or putting yourself in danger by talking to him. It sort of has to be a him, doesn't it?) But we're not morally obligated to do every nice thing that we could do. If we were, we wouldn't have time for anything else.

What Can Gentrifiers Do
Not to Ruin Their New
Neighborhoods?

Around the corner from my apartment, there's a tiny, excellent ital* restaurant and ice-cream shop called Scoops. When the weather is nice, the neighborhood guys hang around all day. It's been there forever. The area has long been Caribbean and African, but people who look like me have moved here in large numbers recently, driving up property values and demand for different, more expensive goods and services. So while Scoops is current on their rent, the landlord has refused to renew their lease and threatened them with eviction. A lot of people have come together to try to Save Scoops. They've petitioned, gotten the attention of the local press, and worked to negotiate with the landlord. We'll have to wait and see whether they succeed.

Many but not all of the potential harms of gentrification are neatly encapsulated in this story. Long-term residents can be displaced (although it's a matter of debate among social scientists how often this happens), they can be priced out of local services, they can lose their feeling of belonging, neighborhoods can become increasingly unequal, and there can be antagonism between long-term and new residents, especially when new residents make frequent appeals to the police.

---

* Rastafarian vegan, more or less.

For the most part, these problems can't be solved through individual action. On her own, an individual newcomer can be friendly with older residents, shop at established businesses, and refuse to rent an expensive apartment or buy an expensive house; unless you're a landlord or a developer or a real estate investor, that's about it. But if you get organized or try to build or leverage institutional power, you can do a lot more. You might help with a petition drive or an appeal to the press, like the Save Scoops campaign, but you can also try to form a tenants union, get involved in local activist groups, or lobby your representatives for changes to housing and commercial rent laws.

That said, a lot of the anger that's directed at gentrification would be better directed at the conditions that make gentrification possible. Neighborhoods are only ripe for gentrification, in general, when the government and private interests have already disinvested in and devalued them. And newcomers only move to gentrifying neighborhoods, in general, because they can't afford the rents or housing prices elsewhere. We could keep real estate prices down by limiting or banning the real estate speculation and usurious lending that inflates them, by legalizing new construction in low-density urban areas (while being careful not to overburden poor and working-class neighborhoods), or, better yet, by massively expanding good quality public housing. We could inoculate neighborhoods against gentrification by funding local public services more fairly and so on. These things are the real problem, not the tattooed guy who works at the new coffee shop.

Should I Give Money to
Homeless People?

If the choice is just between spending the money on yourself and giving it to a local homeless person, it's probably better to give it to the homeless person. Even if they spend some percentage of their money on self-destructive goods like drugs or alcohol (and that's a big if), they'll likely also spend a significant percentage on basic necessities that you already have. The money will do more good for them than it will for you.

But in all likelihood, the choice isn't between spending money on yourself and giving it to a local homeless person. You might have a more or less fixed mental budget for charitable giving, so that the money you give to the homeless person decreases the amount of money you give elsewhere. You might also fall prey to what psychologists call moral licensing, so that giving to the homeless person prevents you from doing other virtuous things, like giving to other charitable causes. If the choice is between giving to the local homeless person and giving to other charitable causes, there's no question that other causes are better.

Consider the Against Malaria Foundation (AMF), an organization that is consistently ranked very highly by charity researchers. For about $4.50, they provide a bed net treated with insecticide to someone in a region where mosquitoes carry malaria. The charity evaluator GiveWell estimates that a donation of $100,000 to AMF would prevent thirty-six deaths—a cost of $2,778 per death.

(This is to say nothing of the number of nonlethal cases of malaria they prevent or any of the downstream benefits of keeping people malaria-free.) By comparison, when the New York City government provides housing for homeless people, because there aren't enough homeless shelters, it often just pays for a hotel room.* The cost of housing a homeless single adult in a New York City hotel room is around $40,000 per year. But even if the city just rented an apartment for them, it's more or less impossible to find a studio for less than $1,400 per month. What's more important: saving someone's life and preventing a bunch of other people from getting malaria or paying someone's rent for a couple of months?

> ✎ Yeah, I get that there are more efficient ways of giving to charity than giving to the homeless. But there's more to charity than efficiency. The homeless person is right in front of me, asking for something face-to-face. And besides, don't we have an obligation to take care of our own communities before worrying about people overseas?

It's true that we feel the needs of people in front of us more acutely than we feel the needs of people around the world that we'll never meet. It's also true that the desire to give back to our own communities—and the sense of reciprocity that backs it up—is very strong. But these impulses push us in a pretty unjust direction if we don't keep them in check. The stronger our special obligations to our own communities, the more those communities will hoard their existing resources. How much educational inequality in the United States is due to wealthy parents donating to their own kids' schools? Giving to local homeless people is, effectively, a variation on that theme.

---

* As I'm writing this, the city is phasing this practice out. But it's still very much alive.

✎ In other words, "Charity: you're doing it wrong."

Yeah, we have to tread lightly here; it's counterproductive to wag your finger at people who are sincerely trying to do the right thing. So what's a better pitch?

> "Charity: nice try, but not quite!"

> "Charity: oooh, you're so close!"

> "Charity: here, let me help you with that!"

What's the Point in
Getting Upset About Things
We Can't Change?

At least a lot of the time, there is none. If you find yourself in a rage because, say, your train is delayed, in a sense you're shooting yourself in the foot. You're already at a loss because you're going to be late wherever you're going. Other things being equal, getting upset makes things worse—and unlike the train it's partially under your control.

But sometimes it does make sense to be upset about things you can't change.

First, maybe you can't change them now, but you can work to prevent similar bad things from happening in the future. If I'm grumpy because I got caught in the rain, maybe I'll be a bit more likely to take an umbrella next time.

Second, being upset about things you can't change is sometimes a matter of self-respect. Suppose I'm routinely singled out by the local mafia for some form of unpleasant but not life-threatening treatment. The mafia is too powerful for me to do anything about the situation by myself, and I'm not interested in the risky and difficult prospect of working with the cops or organizing a movement against them. Now suppose that I feel nothing in particular about this mistreatment. What does this say about me? It's *possible* that I recognize that I'm a victim of some injustice, realize that I can't do anything about it, and have trained myself not to feel resentful or humiliated. But it's also possible that I don't feel anything because I don't think anything especially bad is happening to me—that I'm getting what I deserve, say. Wouldn't this be sad? It sounds like I'm describing a person with no self-respect, someone who doesn't value their own well-being as highly as they value the well-being of others. (James Baldwin described his father as having lost his self-love— and we might say self-respect—because he could no longer be angry at the way white society devalued and mistreated him.) Such a person would be *better off* feeling resentful and humiliated I think, not because resentment and humiliation are intrinsically good for you but because they're closely tied to caring about yourself.

Lastly, emotions are often deeply social things. We feel what we feel, to a great extent, because doing so helps us get along better with other people. So the fact that an emotion is a reaction to something that you can't deal with *by yourself* is often beside the point. If you go through a breakup and you don't know what to do with yourself, telling a friend how sad you are (or just bawling in front of them) will likely encourage them to comfort you, to spend time with you, to reassure you that people care about you. We get sad not in spite of but *because* we can't change our circumstances on our own.

Maybe I should add that there's something suspiciously macho about wanting to master your own feelings, like the next step involves protein powder and some questionable dietary supplements. The impulse to dominate anything—in ourselves or in the world around us—ought to raise some eyebrows. I'll hold on to my Kleenex, thank you very much.

How Do You Cope with the Mortality of Your Own Parents?

Some philosophers might try to take the edge off by trying to convince you that death isn't bad—or at least isn't worth getting upset about—because dead people don't want to be alive, or death is just like sleep, or just like the time before you were born, or because death is necessary to make room for new life, or because there's no point in being afraid of something inevitable, or because your immortal soul will outlive your body (which isn't that great, anyway). I won't do that here.

But there's at least one idea that has been therapeutic for me, so I'll share that. It involves a detour through your relationship to your own mortality. First, notice that death is an all-or-nothing deal: you either exist or you don't. But what *matters* to us about our continued survival is not all-or-nothing.

This is easiest to see using some science fiction-y thought experiments. Imagine that you were about to go into surgery, where 1 percent of your central nervous system was going to be replaced with material from some stranger. One percent of your memories, personality, plans, desires, beliefs, and so on would disappear, and in their place would be the stranger's. How would you feel going into this surgery? Maybe a little scared, a little curious—but I don't think you'd feel like you were going to *die*. But now adjust the thought experiment a bit. What if 25, 50, 75, 99 percent of your central nervous system were going to be replaced? As the number goes up, it feels more and more like this surgery is just murder.

But isn't this really the same thing as aging? Our present selves are products of a sort of gradual fusion of who we were and who we will be. Every day we replace dead cells with new ones. We give up old beliefs and take on new (hopefully, usually better) ones. We lose and gain skills. We forget what happened to make room for memories of what's happening now. I'd be recognizable to my fifteen-year-old self, more or less. But I'd also be a bit of a surprise. It would be sad if I *weren't*.

All of which is to say: in all the ways that matter, for better and for worse, little deathlike things are happening to us all the time. Death is bad for us, but it's not bad in a way that's utterly unlike anything we've lived through before. To the extent that your attitude toward your own mortality is utterly unlike your attitude toward the changes you undergo in your lifetime, it's an uneasy fit.

What does this have to do with your attitude toward your parents' mortality? Well, insofar as your anxiety about their mortality reflects your belief that their deaths will be unprecedented harms to *them*, revising that belief will tend to ease that anxiety. But also, to the extent that your parents (like everyone else) have changed constantly in all the ways that matter, and death is like change within a life, you're more prepared for their deaths than you might think. The problem of coping with your parents' mortality might seem so daunting because it seems so unfamiliar, so monolithic. But you've been dealing with it already, piece by piece, for your entire life.

The question has come up, in one way or another, at a lot of different booths. The first time was at a farmers market in Queens. I forget how we greeted the person who approached the table, but it must have been more or less cheery. She sounded hurt and angry in response. Her mother had just died, and what the hell could we do about it? I think I strung together something halfway coherent about Epicurus and Lucretius,* but I was definitely thrown off. I wish I could go back and have that conversation again—to raise the point about what matters in survival coming in degrees, the importance of focusing on the living, the question of emotional rationality. (What makes an emotional response fit a situation like this? What makes any emotional response fit any situation, for that matter?) But it's hard to talk about this stuff while you're grieving or even while your parents are dying. Better, if possible, to think things through when it's not too urgent.

---

* Epicurus and his later follower Lucretius argued that your death won't harm you. They gave a number of arguments for this conclusion (maybe a bit too many), but the most famous is known as the no-subject argument. When you're alive, you haven't died yet. When you're dead, you won't exist, and nothing can harm you if you don't exist. So your death doesn't harm you when you're alive or when you're dead.

Fun fact: the Yiddish word for a heretic is an *apikores*—an Epicurus.

How Can You Live with
Purpose in Retirement?

I t feels a little ridiculous to say this, writing at the age of thirty-three, but: in a way, this isn't a hard question. Just being retired doesn't prevent you from doing all sorts of purposeful things for other people—volunteering with charitable organizations, supporting friends going through difficult times, giving money (if you have it) to good causes, and so on. If your job, like a lot of jobs, had more to do with making money for your employer than with making the world a better place, perhaps you have *more* opportunities in retirement to do purposeful things for other people. And even if you want to do purposeful things for *yourself* or undertake a project of self-improvement, there are plenty of options—exercise, travel, taking classes, practicing a craft.

But this answer isn't any good, practically, unless you are both motivated to do these things and have the ability to do them. You might not be motivated for any number of reasons. Some people's identities and sense of their value come from their professional status or the sense that their bosses and coworkers and clients *need* them. Some people are forced into retirement against their will, which can be a depressing and embittering experience. Some people are paralyzed by their fear of death or by internalizing noxious or debilitating attitudes about what aging means. Some people just never cultivated the habits of curiosity or intrinsic motivation or whatever to get out of the house in the morning. And while there is evidence that we underestimate the "productivity" of older people, you might not have the ability to do a lot of what you would like to do, for others or for yourself.

There's no one-size-fits-all way of responding to this philosophically; after all, these aren't entirely philosophical problems. But one thing philosophy can do is to help articulate the distinctive values that come with the experience of retirement and the experience of aging, more generally. It's hard to pull this off without overgeneralizing, pandering, or waxing sentimental, but for people who are struggling with the feeling of not being needed anymore or not knowing what role they have to play in the world around them, it might be helpful.

One thought: we are all more or less vulnerable and dependent on other people all of the time. This shared vulnerability is, in part, a strength—it can serve as a basis for solidarity, sympathy, and reciprocity. But it's hard to keep this vulnerability in mind, as long as we're healthy, working adults—less reliant, some of us like to think, than relied upon. This is one way that the experience of aging and retirement might be valuable: it might force you—and the people around you who are paying attention—to appreciate the significance of our vulnerability. And as our own mortality becomes more and more salient, the full extent of others' dependency on us can also come into clearer view. If we are sad when someone dies, or anxious at the prospect of their death, it's not generally because of their contribution to the formal economy. We don't just depend on people for their jobs.

Another thought: the philosopher-psychologist Karl Jaspers claimed that our worldviews (which I take to be the general beliefs and attitudes that we hold most closely and draw upon most extensively) are held in place by "limits"—possibilities that we refuse to consider, contradictions we refuse to resolve. But every now and again, we find ourselves confronted with those limits. We experience (maybe anxiously, maybe joyously, maybe some other way) the inadequacy of our basic understanding of how the world works, our place in it, and what matters. Jaspers called these experiences "limit situations." We can encounter limit situations at any point in our lives, but retirement itself is a limit situation for many people, and aging in general comes with all sorts of limit-testing challenges and discoveries. I'm far from certain, but I hope that we learn from our limit situations cumulatively, that we get more imaginative and competent at dealing with them over time. If this is right, we might hope for just a bit more, that this acquired wisdom is transmissible by word and example. I'm not retired and I'm still young-ish, but the more I think about this, the more I think there is to keep an eye out for.

What Makes Someone
Mentally Ill?

We could approach this question by just gathering the most uncontroversial examples of mental illnesses, describing what they have in common, and describing what makes them different from mere quirks or idiosyncrasies. But I don't think that's the most useful tack to take. First, even the most uncontroversial examples of mental illness are still pretty controversial, since a lot of people are deeply skeptical of the way we talk about mental illness—antipsychiatry theorists, disability rights activists, and other people who suspect that we sometimes try to medicate our way out of big-picture societal problems. It would be nice to have an account of mental illness that helps answer these skeptical concerns and doesn't just assume they're wrong at the outset. Second, the American Psychiatric Association's *Diagnostic and Statistical Manual of Mental Disorders* changes pretty dramatically from one edition to the next. To the extent that this represents the professional consensus about what counts as a mental illness, it's a bit of a moving target. But it would be nice to have an account of what mental illness is that could stick around for a while—indeed, something we could use to figure out what the *DSM* gets right or wrong. And third, we use the concept of mental illness to do all sorts of things, practically and theoretically. If we try to figure out what mental illness is before figuring out what we use the concept of mental illness to do, it's not clear that the account will actually help us do those things.

Luckily, there's another approach—what the philosopher Sally Haslanger calls ameliorative analysis, which is more or less what the philosopher Rudolf Carnap called explication. First you ask what's the point or purpose or function of the concept of mental illness. What do people do with the concept of mental illness that explains why we keep the concept around? Then you ask what would best fulfill that purpose. If the concept of mental illness is a sort of tool, how can we engineer it so that it does its job(s) better?

If you ask me,* the concept of mental illness has at least three functions. One function is to constrain the decisions of research psychiatrists and psychopathologists. They study the nature, causes, and treatment of mental illness, so judgments about what is or isn't a mental illness affect how they spend their time and attention.† Another function is to guide decisions about medical care. People seeking care for mental illness are referred to psychiatrists, clinical psychologists, and mental health social workers, not to other types of care workers. Lastly, mental illness serves a moral and legal function. If we think that a behavior is caused by mental illness, we don't hold people accountable for that behavior in the way we normally would. This is vague, but I think you know what I have in mind; when we think someone's bad behavior is caused by mental illness, we're more likely to seek medical care for them, less likely to seek retribution, and less likely to experience, or fully endorse, certain emotional re-actions to the behavior—e.g., resentment or disgust if it's someone else's behavior, guilt or shame if it's our own. (Importantly, someone might agree that these are the functions of the concept of mental ill-ness and still think that there is no such thing as mental illness. You might think that there is nothing for psychopathologists to study, no legitimate medical role for mental healthcare providers, and so on.)

---

* For some reason.

† It would probably be more accurate to use "disorder" rather than "illness" here. The initial question was framed in terms of mental illness, but feel free to substitute in "disorder" wherever you like.

So the question is: How can we define the concept of mental illness so that it best serves these three functions? Or, if each function is best served by its own definition of mental illness, how should we define the concepts of mental illness?

And here's where I run out of answers. I don't know enough about the methodological strengths and weaknesses of psychopathology, the track records of various sorts of mental health interventions, the likely outcomes of different approaches to mental illness in criminal law, and when people are benefited or harmed by treating bad behavior as a medical problem. (Sorry!) But even if we can't say what counts as a mental illness, in general, we can ask in any particular case whether this is something psychopathologists should study, something mental health workers should treat, or something we should address using the tools of social control that morality and the law typically afford us. I'll say this, though: researchers estimate that around 20 percent of people in US jails and prisons have "serious" mental illnesses. The majority of incarcerated people are rearrested within a few years of getting out. The cops have a decidedly murky track record at handling people who have a history of mental illnesses. (Deborah Danner, Saheed Vassell, Sandra Bland, Charles Kinsey, Kwesi Ashun . . .) So however we define "mental illness," there's good reason to think that many of the problems currently treated by the criminal justice system would be better treated by mental health workers.

When You Buy Something from a Poor Country, Are You Exploiting Their Workers?

Here's one way of thinking about exploitation, at least for economic purposes: I exploit you when I should be blamed for overcharging or underpaying you for some good or service.

I've added the "should be blamed" because I (and the questioner, I bet) think of economic exploitation as a moral concept. If, for example, I overcharge you for some trinket because we're both blamelessly ignorant of some defect that renders it worthless, it's a stretch to say that I've exploited you.

The trickier thing is to say what counts as overcharging or underpaying or what counts as a fair price. But here's a theory that works pretty well. It ~~shamelessly appropriates~~ draws on the philosopher John Rawls's idea of the veil of ignorance. Suppose that you want to buy a widget from me. Now ask: What if we had all the information we needed to reason about the transaction? What if you and I knew all the relevant facts about the quality and manufacture of the widget, the prices widgets generally command, each of our needs and preferences and financial situations, the psychology of consumer behavior, and so on? But here's the catch: the one thing we don't know is when the transaction is over, who is going to be in the position of the buyer and who is going to be the seller. The veil of ignorance is this imaginary bargaining situation. And the theory is: the price we would agree on behind the veil is the fair price for the transaction.

A few wrinkles:
- The theory would need to be clarified/modified to account for things like buying lottery tickets and buying information, where a certain sort of ignorance is intrinsic to the transaction. I don't know how to do this, but it's not really relevant to the question at hand.

- The theory would also need to be changed to deal with the exploitation of children and nonhuman animals.

- Slavery is a paradigm case of exploitation, but it doesn't fit neatly into this picture, because, in a sense, slave owners don't transact directly with the slaves who work for them. I think the answer is to take an appropriately expansive conception of transaction. Slave owners transact with slaves at least in the sense that if the slave did agree to work for the slave owner, the slave owner should pay them for that work.

Of course, it's impossible for us to actually go behind the veil of ignorance. But we can estimate what we would do behind the veil in various ways—through our imaginations, by getting people who have been both buyers and sellers to estimate what they would take to be a fair price, by averaging the going rates in markets that are favorable to buyers with markets that are favorable to sellers, by averaging the prices the buyer and seller actually say they would be willing to pay and then adjusting from there in light of specific asymmetries in bargaining power and relevant knowledge, and so on.

I like this way of thinking about exploitation and fair prices because it substantiates and explains what a lot of us find fishy about people agreeing to transactions under duress. It can explain why, for example, it's typically exploitative to pay people far below the minimum wage, even when they're willing to do the work (because the employer can only get away with such low wages by taking advantage of their outsized bargaining power). It also explains what counts as price gouging (rather than natural price fluctuations in response to scarcity) and why price gouging is generally wrong (because the seller wouldn't want to buy the good for the gouge-y price if they were in the buyer's position). More generally, the account strikes a balance between the principle that fair prices depend on the individual preferences of the buyer and seller, and the principle that unfair prices arise from ignorance and lopsided distributions of bargaining power.

To get back to the question, though: on this account, you can't exploit someone that you don't transact with directly. So unless you're buying something from an overseas worker directly, you're not exploiting them.

But we can't get off that easy. After all, if an overseas worker is being exploited by their employer, and I go to the store and buy something produced, in part, by that worker, I am paying someone to pay someone to pay someone . . . to exploit that worker.* And typically, if it's wrong to do something, it's wrong to pay someone to do it. So it seems that if I find myself at one end of a supply chain and exploitation at the other, then I'm probably doing something wrong.

Maybe. But there are two features of the consumer situation that can take some blame off your shoulders. One is that consumers make decisions under what psychologists call bounded rationality. We could always make better choices but doing so would sometimes require time or brainpower that we would be better off using else-

---

* In a sense, I'm not paying someone to pay someone . . . to exploit the worker, because people give manufacturers their money in exchange for actual goods, not in exchange for exploiting their workers. But is this a distinction that matters? There is a tradition in philosophy and the law according to which there is a morally significant difference between the intended effects of our actions and their unintended but foreseen effects. This is called the doctrine of double effect. The question is whether the doctrine of double effect holds in general or in the case of overseas exploitation in particular. But since the entire purpose of the doctrine of double effect, as far as I can tell, is to make excuses for people who knowingly harm others, it's not something I take very seriously.

where. Perhaps there are ten brands of shampoo on sale, only one of which is exploitation-free, but it would take me several hours of research to figure out which brand is the right one. If avoiding the consumption of exploitative products would require this sort of exercise of time or brainpower, then blaming you for nonoptimal consumer choices seems like a nonstarter. (It's easy to push this point too far, though. "How was I supposed to know that Evil Corp. is evil?")

The other is that sometimes we don't have meaningfully different choices between consumer goods. You need to eat vegetables, and it could be that all of the vegetables available to you (or all of the vegetables available to you without a goofy expenditure of time or money) are harvested by underpaid workers. So while you don't really have a choice in the matter, maybe their employers (or the vegetable-growing industry as a whole) do. In general, when you see a discussion of the ethics of individual consumer behavior, it's helpful to ask whether we would be better off talking about employers or regulators or industry norms instead.

Is It OK to Have a Pet Fish?

Depending on the fish, if you take proper care of it, I think the answer is probably yes. The question for me is: does the net pleasure (no pun intended) a fish enjoys in captivity over its lifetime, plus the net pleasure the owners derive from having the fish, exceed the net pleasure a fish enjoys over its lifetime in the wild?

Part of why this question is so hard is that fish are so inscrutable. We can tell how well our pet mammals are doing at home because we are, more or less, them. But fish are pretty alien creatures. It's hard to know how well they're doing because it's hard to know if they are even capable of doing well or poorly. So let's suppose for the sake of argument that pet fish are capable of experiencing pain and pleasure. They must feel pain at certain things that tend to make it harder for them to survive and reproduce and pleasure at things that do the opposite. The question is which of those things actually cause them pain and pleasure and how much?

Take the common goldfish (and its wild counterpart, the Prussian carp). Goldfish enjoy certain benefits raised in captivity. Most obviously, they don't have to worry about predators or running out of food.

Now, there are a lot of possible causes of unpleasantness in a pet goldfish's life. If a goldfish is kept alone in a fishbowl, it won't get enough oxygen, it will drown in its own waste, it won't get the stimulation it needs from socializing, and its visual system might not work because of the shape of the bowl. It will almost certainly die younger than the average Prussian carp. But all of these problems can be solved by taking better care of the fish—just keep it in a big enough, properly maintained, rectangular tank (or, better yet, a pond) with at least one other goldfish. A well-cared-for goldfish will live at least as long as a Prussian carp on average.

It still won't be able to perform some of its natural behaviors, from which it might derive some pleasure. In particular, it won't be able to forage, it won't be able to mate (unless opposite-sex goldfish are paired, unwisely), and it won't be able to migrate. (Apparently, wild Prussian carp have been observed migrating more than fifty miles.) Now, most of a Prussian carp's diet isn't very hard to catch, so it's not clear why they would derive much pleasure from foraging (rather than just eating). But missing out on mating and migration seem like potentially significant losses. Perhaps some of this could be mitigated by redecorating the fish tank from time to time. Still, it seems to me that this unpleasantness is probably outweighed by being kept safe and well fed. After all, a fish that preferred migrating and mating to eating and avoiding predators wouldn't stick around for very long. Not to mention the fact that people who keep pet fish presumably enjoy having them around.

But isn't keeping pets basically slavery?

Keeping pets is definitely like slavery in some respects. You prevent them from moving around freely and from deciding for themselves the basic form of life they're going to have, and you don't pay them for the service they provide you. But while there are a lot of ways of explaining what makes it wrong to enslave human beings, those explanations tend not to carry over well to nonhuman animals. Paying them for their work is beside the point. They aren't, in general, visibly pained by having many of their decisions made for them. People care about (and are capable of) planning their lives in ways that nonhuman animals can't. When pets escape from being pets, they aren't generally better off. And so on.

So on balance, a well-kept goldfish probably brings more joy to the world than your typical Prussian carp. If you want one, knock yourself out. But you know, there are these things called dogs.

When Should We Trust
the Experts?

When I present this sort of question in class, I can rely on someone (usually a few someones) to read, more or less, off the following script: you should *never* trust teachers/journalists/doctors/scientists/whoever. They don't know what they're talking about. They're trying to sell you something. For any expert who tells you one thing, you can find another saying the opposite. They just want your unquestioning obedience. They're paid off by the man. Belief in the experts is the mark of the rube or the conformist—or at least the mark of someone who can't or won't think for themselves.

This script is sort of in the air, but it seems to float completely free from how people *actually* relate to experts. Mrs. Rosenblum was, I suppose, paid by the man to teach my third-grade math class long division, but she wasn't just making stuff up. If you want to find out the weather or local traffic conditions, you go to the news (or whatever source has been deemed reliable by Google). Few of us can explain the evidence for the conclusion that the earth goes around the sun and not vice versa, but we all know it about as well as we know anything. To refuse to believe what the experts tell us about these things wouldn't be a sign of an exceptionally clear, critical view of our information ecology. It would just be paranoid.

But of course the script gets something right. We shouldn't *always* trust the experts—or at least people who purport to be experts or have expert-ish reputations or credentials. Before Copernicus, the unanimous scholarly consensus (in Europe, anyway) was that the earth is the center of the universe. The press gets things wrong all the time, sometimes spectacularly, sometimes systematically. Some whole sciences might turn out to be mistaken—disciplines that keep going by discrediting anyone who disagrees with some basic assumptions, say. And of course the experts sometimes disagree with one another.

One place to start in thinking about this stuff is to ask why you believe what anyone tells you at all. Some philosophers have argued that the mere fact that someone tells you something (in the absence of any special contradictory evidence) justifies you in believing them. Maybe, but I don't think we need to go that far. Suppose a coworker tells you that they were assigned to complete some report. Why trust them? I can think of a few reasons:

- People, generally, know pretty well what they have or haven't been told to do at work, especially if they're confident enough in what they know to talk about it out loud. So your coworker probably knows what they're talking about.

- It might be hard to think of any reason they would have to lie, and there's plenty of reason for them not to. You will likely hold them to account if they do, and your relationship with them will likely be a little frayed. Most people don't want to take that risk.

- Sometimes you can tell when someone is lying, not so much from what they say as how they say it. If you have a reasonably well-calibrated lie detector, and it doesn't go off, that's a good sign.

- Even if you're not in a position to detect a lie, your coworker presumably isn't a psychopath. Most people feel guilty about lying about this sort of thing, especially if they don't have a good excuse.

+ And of course your coworker has likely told you other stuff in the past that you've been able to verify for yourself. If they were right then, they're probably right now.

You can imagine other speech situations that differ from this in some important ways—an exchange with a stranger on a YouTube comments thread, say—where you might not trust the speaker quite as much. But at least when people are talking about prosaic or verifiable matters of fact in the context of valuable, ongoing relationships whose continued existence depends on trust, you can generally be pretty confident that they're right. (It's a good thing, too, considering how little I know on the basis of my own direct observation.)

That's just ordinary conversational trust, though. Our deference to experts goes beyond that. We believe what they say to an unusually high degree; we give them platforms for expressing their views that we don't give to just anyone; we give expert opinion special weight in our political decision making through institutions like the Congressional Budget Office and the Census Bureau. If this sort of deference is justified, it will take more than whatever reason we have to believe a passing remark in a workplace conversation.

But at least some of the time we do have special reasons to defer to experts. Whether we ought to defer to an expert in any given case is a matter of whether these reasons obtain. Here's a nonexhaustive list:

+ *Credentials:* Experts might have credentials—fancy jobs, academic degrees, platforms. It can be hard for a layperson to tell the difference between a legitimate credential and a meaningless one, but one good test is whether those credentials are recognized by a wide range of people with different sorts of credentials.

- *Fact-Checking, Debunking, and Accountability:* Experts might work in a field where what they say is either rigorously fact-checked by some editor or third party or where there are strong incentives for debunking false claims after the fact. Experts might suffer some harm to their reputation or their professional standing if they wind up on the wrong side of a fact-check or a debunking. Compare a political pundit who is never called out on their past false predictions with a scientific field that regularly publishes failures to reproduce earlier experiments.

- *Consensus:* Sometimes all of the experts agree with one another or they communicate through some document that represents a community consensus. But even when the experts don't agree, there might be strength in numbers. If an expert speaks on behalf of a majority of experts in a field, that's a reason to defer to them, especially or at least if there's some indication that this majority isn't just the result of some groupthink or echo chamber or something.

- *Funding:* Some experts are funded by universities or public grants or private enterprises who have a market-based incentive to get at the truth about some matter. Others are funded by people and organizations that are governed more by self-interest or profit than concern for the truth— psychiatric conferences funded by the pharmaceutical industry, libertarian think tanks funded by ghoulish billionaires, educational programs funded by fundamentalist religious groups, nutritionists selling some diet products, expert witnesses paid for by the prosecution, practicing doctors who get their information about the efficacy of new drugs from the manufacturers. These things are matters of degree, not of kind, but some funding sources lower the credibility of a speaker more than others.

None of these reasons to defer to experts are perfectly reliable. And a lot of the time, we won't be able to tell whether they obtain without doing extra research. But that's not so much a philosophical problem as a problem with how information actually spreads in our time.

How Can We Raise Children
to Be Good People?

A kid's school, peer group, genes, and the dominant culture place some pretty strict limits on what their parents can do to shape their behavior. But of course parents have *some* influence on how their kids turn out. Here are a few suggestions for how to make better people that enjoy a reasonable amount of empirical support:

- *Give kids admirable, relatable role models.* Kids can be motivated by stories about people (especially people who are only a couple of years older than they are) who do morally extraordinary things or who overcome obstacles or prevailing norms to do what's right. In an often immoral world, good people are rebels; everyone loves a rebel.

- *Make abstract principles concrete and emotionally engaging.* Saying "be fair" or "be kind to others" is good advice, but it doesn't affect people emotionally. People are motivated to act by what affects them, and we're affected by particular, detailed examples more than by general or quantitative descriptions of things.

- *Make it easier to be moral.* To whatever extent you can, "nudge" kids in the right direction without materially rewarding them for good behavior: require them to opt in to (rather than opt out of) bad choices, frame good choices positively, avoid situations where figuring out the right choice will be annoyingly brainy. There's probably some room for material rewards, too, but that's riskier. The more extrinsically motivated the child, the flimsier their moral habits.

+ *Set rules that the people around the child actually follow.* Telling someone that it's against the rules to do such and such and then putting them in situations where people conspicuously do such and such is a very good way to get them to think that the rules aren't supposed to be taken seriously.

I'll add a couple of other suggestions, for which I don't have any kind of documented or systematic empirical evidence but that I think are important:

+ *Encourage moral numeracy.* It's really hard for people to take quantitative information seriously when they're reasoning about moral problems. But without this ability to think quantitatively about morality, we can't keep our moral attention where it belongs. I don't know the best way to help kids figure this out but giving them practice couldn't hurt. If your kid wants to give to charity, ask them to justify their charitable giving in numerical terms.* If your kid has to do a research project on a social problem of their choice, ask them for some numbers that show that their social problem is more important than some possible alternatives. If your kid thinks that some division of labor or resources at home is unfair, ask for a way of quantifying that unfairness.

---

* Figuring out a way to do this that doesn't inhibit their charitable impulses might not be easy. Maybe start by praising them for their charitable impulses and guide their impulses in a more useful direction once they're firmly in place.

In general, when we think about moral education, we tend to focus on developing good habits for dealing well with the people in our immediate environment—politeness, respect, and so on. But while being polite and respectful in conversation is better than the alternative, this doesn't have that much to do with solving the biggest problems in the world today. To that end, I think it's important to:

+ *Encourage kids to participate in collective action.* Political corruption, climate change, poverty and economic inequality, the horrors of animal agriculture, and so on aren't going to be solved by individuals putting around doing their own thing. They're going to be solved, in large part, by organizing, lobbying, regulation, boycotts, strikes, and movements to change prevailing norms. Since these are moral problems, their solutions are our collective moral obligation. Of course, kids can't fully participate in most meaningful forms of collective action, and even if they could, a lot of adults wouldn't take them seriously. But we can create opportunities for them to get some practice, from student government to park cleanup projects to participatory budgeting. I don't know what sorts of interventions make kids more likely to participate in important forms of collective action as adults but getting them acquainted with the power of collective action firsthand seems like a step in the right direction.

What Makes a Word Sexist?

A pretty good answer: a word is sexist if, in general, its use promotes sexism—that is, unjust discrimination against or oppression of people on the basis of sex or gender. One type of word that can do this is what philosophers call a thick evaluative term. Thin terms, such as "good" or "should," can be used to evaluate pretty much anything. They don't tell us much about *what* we're valuing. Thick terms, such as "courage" or "gluttonous," are different. If I call someone courageous, I'm not only praising them, but I'm also telling you a bit about what I'm praising them *for*—something to do with the ability to act in the face of danger. Some thick terms are gendered, in the sense that they're used more or less exclusively for one gender or another. Now, the fact that a term is gendered doesn't necessarily make it sexist. If, for example, men were really more likely than women to explain things in a tone-deaf or condescending way, and the rise of the word "mansplain" drew attention to that fact, then it would seem that there's nothing unjustly discriminatory about it. But if a thick term has no excuse for being gendered, then it's sexist. "Slut" is a good example. It's almost exclusively applied to women and girls. But more importantly, there is no counterpart gendered term for men and boys, not because men and boys aren't sexually promiscuous but because we don't judge them for being sexually promiscuous. So the use of "slut" promotes unjust discrimination against women on the basis of sexual promiscuity. You can think of all sorts of examples of sexist thick terms, on this pattern: "fairy," "nag," "dowdy," "cow."

A slightly better answer: a word is sexist if, in general, its use promotes sexism *or* the attitudes it is used to express are sexist—that is, they are attitudes in favor of sexist practices. I want to leave open the latter possibility because there might be cases in which the effects of using some word are too diffuse and obscure to pin down, but we want to hold people who use the word to account because of what this use says about *them*. To take a made-up example, suppose there is a slur against women that only committed misogynists use in their journals or in private conversation with one another. It's doubtful that this slur by itself causes much harm because of its restricted audience. But then suppose that I find that a friend has used this word in some private message. Even though the word is harmless, I would still be within my rights in asking my friend to explain himself.

That said, this is a question about words. But words are pretty abstract things. In order to figure out whether a *word* is sexist, we have to generalize across all of the occasions in which it's used, most of which we'll never observe. It's a lot easier to figure out whether a particular utterance or speech act is sexist. Take a word like "bitch." It's pretty much hopeless to say whether the word itself is sexist, since it's used in so many different ways. But if we look at particular utterances of the word, things come into focus. When Lizzo sings "I'm 100 percent that bitch," she's not expressing a sexist attitude, and she's not promoting any sexist practice. But suppose I ask a woman on a date and she declines. It would be sexist to call her a bitch. Or take "Mrs." It's often completely harmless to refer to someone as Mrs. So-and-So; this might just be what they like to be called. But suppose I create a job application form that requires people submitting it to check a box for "Ms.," "Mrs.," or "Mr." Then I'm requiring women—and not men—to reveal potentially damaging information about their marital status. A lot of what we might think of as borderline or questionably sexist words can be better understood by getting down to cases. "Pro tip: if you're ever stuck on a question about the meaning of a word or phrase, change the topic to what people actually do when they use that word or phrase."

What Can You Do When Nothing You Do Matters?

The person who raised this question was talking about climate change, but the problem is even bigger than that. A lot of our moral and political problems revolve around things you might do that don't matter: voting in big elections, littering, paying your taxes, light shoplifting, microaggressions, moving into a relatively expensive apartment in an affordable neighborhood, small charitable donations, grazing a farm animal on public land, fishing. Or more precisely, it doesn't matter if *you* do them. They make a very big difference if lots of people do them. If you litter or shoplift a piece of candy or don't pay your taxes, the world won't stop turning. But if everyone litters and shoplifts and dodges their taxes, the streets will fill with waste, the stores will close, and the government will collapse.

Philosophers call these collective action problems. One way of putting the problem is: if it would be very good for a large number of people to act in a certain way, but it doesn't make much of a difference whether any one person acts that way, how do we get people to act that way?

Some solutions are legal or institutional. We fine people large amounts of money for tiny acts of shoplifting; we regulate who can fish, in what season, and when they can keep what they catch; your workplace might lay out a no-tolerance policy for certain types of microaggression. Anytime you have an institution that is in a position to change the relevant incentives for a large number of people, you have a potential (partial) solution to a collective action problem.

These solutions have their limits, though. First, there are some elements of our personal lives that we just don't want lumbering, ham-fisted institutions to get involved in. We *could* have the cops write tickets for microaggressions, but I'd rather we didn't. Second, in some circumstances, legal solutions only work when they're enforced widely enough. If my town outlaws fishing at certain times of year, and the next town up the river doesn't, all of the fisherpeople will just go there to fish. A single manufacturer that opts out of a profit-making polluting practice might stay afloat by milking it for good PR, but they might just go out of business. This is one of the special problems that climate change poses: to the extent that climate change is a global phenomenon and the legal solutions to it would have to be enforced internationally, we'd need an effective international government to do so. The problem is that we don't have one. And lastly, even if, say, the United States and China could singlehandedly pass laws that would halt climate change, they might not want to—both because very rich people stand to make a lot of money from a melting planet and because (correctly or not) unilaterally giving up some profitable enterprise in the interest of fairness makes people feel like chumps.*

When legal or institutional solutions don't work, moral solutions might. We keep littering under control largely by sustaining a moral norm against littering. We teach kids not to litter in schools. We invent new moral vocabulary ("litterbug"), tailor-made for the purpose of keeping litter off the streets. We call people out for littering in public. We build identities around clean living. We create an expectation that not littering is just part of what it takes to keep up, morally, with the Joneses. In that way, people either internalize the moral

---

* To be clear, I don't want to suggest that legal solutions to climate change are hopeless—only that they have some serious obstacles to overcome.

judgment that littering is wrong, or they don't litter because they're afraid of what other people will think.

Again, these solutions have limits. First, moral norms are hard to change. The arguments against eating factory-farmed meat are absolutely overwhelming. If you are even a little familiar with the basic facts on the ground (about the treatment of farmed animals, the environmental impact of animal agriculture, the risk to human public health posed by waste lagoons and farm-grown antibiotic resistant bacteria, the corrupting influence of the animal agriculture lobby on state and federal government, the treatment of workers in the industry), you have to delude yourself in order to think that it's not a terrible mistake. But people like eating meat, and their neighbors eat meat, and that's about all it takes to keep the practice going. Second, moral norms only motivate us so much. See Milgram's famous obedience to authority experiments (which show that people are willing to inflict extreme harm because someone in a lab coat tells them to), or Darley and Batson's slightly less-famous Good Samaritan study (which shows that whether people decide to help someone in extreme distress is sometimes just a matter of whether they're running late for a trivial appointment), or, I'm sure, your own rich firsthand knowledge that people don't always do what they believe to be right. And third, at least when it comes to collective action problems that involve the economy, moral solutions have, at least in the past, tended to focus on individual consumer behavior. It's on each of us to keep track of what we eat, how we heat our homes, how often we travel, when to carry a reusable grocery bag or a metal straw, and so on. Of course it would be good if we did these things. But this might be a distraction both from supply-side morality—why are companies selling us all this crap, anyway?—and deeper structural transformations in the economy that might change the scope of our consumer decisions altogether.

Maybe all of the above is kind of obvious. But if it is, it's worth wondering why so many intelligent adults seem not to get it.

A lot of people who stop by the booth don't have a question or aren't really looking for a conversation. They just have a theory or a story they want to get off their chests. This is 100 percent fine by me. If the booth is a place for people to express themselves in a way they can't elsewhere, that's great. But sometimes what start out as monologues turn into real conversations. One lady told us, in so many words, that climate change had made her more selfish. She can't do anything to solve the problem herself, and so there's no point in getting worked up about it. Better to just focus on living her own life. As she was saying this, a passerby stopped to voice her disagreement. Maybe your actions alone don't matter, but we can fight climate change together. We had a lively discussion of free riding and collective action. As one woman left, she picked out a lollipop from the bowl of candy that we put out.

"I think I've earned this."

I responded, "Yeah, I think you've given us about twenty-five cents worth of philosophy."

Without missing a beat, she said, "Well, I've certainly given you my two cents."

Should We Live
in the Present?

There are a lot of different things this might mean, but one is illustrated by the famous marshmallow test: offer a kid one marshmallow now or two marshmallows after a few minutes of waiting and see what they do. "Living in the present" here means preferring a smaller present reward to a greater reward in the future. We usually say that the kid who lives in the present, who opts for the single marshmallow, is screwing up: they're acting impatient or shortsighted or weak willed.

I don't disagree when it comes to the marshmallow test. But that's a kind of extreme case. Doesn't it *sometimes* make sense to prefer smaller present rewards to greater future rewards? Philosophers and economists call this sort of preference for smaller present or near-term rewards to larger future rewards *temporal discounting*. People do, in fact, discount the future, although different people discount it to different degrees. But the philosophical question is whether it's rational to do so.*

I tend to think it is. Suppose that a genie comes to me and offers the following choice. I can have one marshmallow today or a lifetime supply of marshmallows starting in twenty years. The genie informs me that I will be alive in twenty years, that I will still enjoy marshmallows as much as I do today, that I will have the willpower not to overindulge, and that I will live long enough to enjoy the lifetime supply for quite a while. At first, the choice is a no-brainer; a lifetime supply of marshmallows is too good to pass up. But, the genie tells me, there's a catch: in twenty years, I will be a complete jerk. I will abandon all of the people and things I care about most. I will spend all my time on the internet, arguing with strangers about the Marvel Comics Universe.

It's hard to know what choice I would make in this situation, but I could imagine myself saying something like: "Oh, future Ian is a jerk. I guess I wish him the best, in the sort of disinterested way in which I wish everyone the best. But I'm not going to give up *my* marshmallows for him. I certainly don't want to reward him for his bad decisions."

In any case, this situation is just an exaggerated version of something that happens to many of us in the real world. We change, often in ways our younger selves would disapprove of. These changes

---

* We should note that even opponents of temporal discounting will accept it in a qualified way. One qualification has to do with certainty. If you are choosing between two options, one of which you are 100 percent confident will lead to some present reward and the other of which you are only 40 percent confident will lead to a future reward that's twice as nice, people generally agree that it makes sense to go with option one. Another qualification has to do with the rewards themselves. Some rewards, like eating a delicious marshmallow, are enjoyed more or less in an instant. Other rewards, like getting a nice new house, you enjoy more the longer you get to have them. When we're asking whether temporal discounting is rational, we're asking whether it's rational (a) adjusting for uncertainty and (b) when the rewards in question are instantaneous.

are hard to notice day to day, but that's how they unfold. If what's in my own interest isn't just whatever is in the long-term of interest of whoever I happen to become but also what is in the interest of me *as I am right now*, then temporal discounting might reflect my best interests.

Another argument for temporal discounting has to do with the relationships between our preferences and our time-sensitive emotions. Some of our emotions are time sensitive in a certain way. We are afraid more or less only of what is in the future.* We regret only what is in the past, and so on. Our preferences and our emotions are tied together in a certain hard-to-specify way. But if fear, for example, is a future-directed emotion, and it's rational for fear to lower our preferences, it's sometimes rational for us to disprefer something in the future to something similar in the past or the present.

This isn't an open-and-shut case. There are good arguments against temporal discounting: that it makes some of our preferences arbitrary, that it opens us up to clever sequences of bets that are guaranteed losses. But I'll let you work out those arguments for yourself.

---

* There are some counterexamples to this. I might know that a potentially deadly surgical operation has been completed and still reasonably be afraid that it didn't go well. But this is the exception that proves the rule.

PART III

Questions You Didn't Know You Needed Answered

Is Ketchup a Smoothie?

Part of what I love about this question is that it does what a lot of great philosophy does: it takes a familiar, otherwise unremarkable feature of everyday life—in this case the smoothie—and makes it seem unfamiliar or puzzling. It's really hard to say what we mean by "smoothie" in a way that excludes ketchup but includes all of the uncontroversial examples of smoothies. But of course, if no one had ever put the question to you, you would never have included ketchup on a list of smoothies.

I think the answer lies in the subtle pressures that shape how we use a word from one context to the next. Think of the word "weight." Your physics textbook tells you that "weight" means *impressed gravitational force*. But even though astronauts in orbit around earth are under considerable impressed gravitational force, we still call them weightless, roughly because if you put one on a normal scale it wouldn't budge. So in contexts where the textbook definition is in charge, we'll use "weight" to mean *impressed gravitational force*. But in contexts where it's important that we're using normal scales to measure weight, we'll use "weight" to mean *whatever normal scales measure*.

The ketchup question gets us to look for a single definition that will include all the smoothies and exclude all the nonsmoothies. This desire for a definition will push us to use the word with an unusual—and unusually uniform—meaning, not what we mean when we put it on diner menus, for example, where our goal is to communicate reasonably clear expectations about what you will get when you order something. So the answer is: maybe ketchup is a smoothie now, but it wouldn't have been if you hadn't asked the question. Either way, if you drink it through a straw, please seek medical attention.

> A woman came up to us with her daughter, who looked to be about nine. The mom asked, "Do you have a question?" The daughter thought for a second, then gave us the smoothie question. Before we could say anything, the mom objected, "No, that's not a philosophical question." But that's exactly what it is.

If Humans Colonize Mars, Who Should Own the Land?

Let's take a stab at it.

First, what does ownership mean here? Very roughly, you own something if you can within the bounds of the law do whatever you want with it, and you can seek redress from someone if they use it without your permission. So questions about who should own something are, roughly, questions about who should be authorized to make decisions about its use.

There might be physical or practical limits on the sorts of uses we could make of Martian land, so that it would only work as, say, a mine or a scientific research field or a public park. Different sorts of limits would call for different owners, or at least stewards. But let's suppose that Mars is going to be used in all the ways that land on earth is used. People will live, work, and play there. They'll make useful things out of the land and take up space as they go about their affairs.

A bunch of different groups would have plausible claims to the land. The first occupants of the land could say that it's theirs, just as explorers on earth have done in the past. Whatever governments or corporations paid to send the occupants there could say that they deserve it as a return on their investment. The people who actually work to turn the land into something useful could claim that since they've created something valuable the value should accrue to them.

Each of these claims runs into its own problems, both conceptually and politically. (Conceptually: Would the first occupants of Mars own the whole planet or just the paths their bodies cut across its surface? Wouldn't the conversion of Martian land into something useful result, in a sense, from the labor of everyone who has ever lived? Politically: Since the richest governments or corporations will be the first to get people to Mars, won't Martian colonization just deepen economic inequality?) But they all have one fishy thing in common: they're all backward facing. Instead of justifying a distribution of property on the basis of how it will likely work out in the future, they appeal to some right grounded in the past. But the past is past. I want things to work out going forward. Now, maybe these distributions *could* be justified in terms of their consequences, but why not think about how to achieve the best consequences directly?

Well, figuring out which land ownership scheme will lead to the best consequences requires us both to dig into some risky speculation and to settle the difficult question of what makes one outcome better than another. It's above my pay grade.

Still, there's at least one pretty good rule of thumb to help get the ball rolling: other things being equal, people should have a say in a decision proportional to the decision's effects on them. For example, if a group of people are planning to eat dinner together, the lion's share of the deliberation about what they're going to eat should fall to them, because the decision primarily affects them.* Of course, we aren't always great judges of what's in our own best interest, but when an indefinitely wide range of decisions is to be made, you are likely in as good a posi-

---

* It's worth thinking this through in some more detail. The restaurant they go to will have a say in their meal by setting prices and by setting limits on what sorts of things they're willing to cook, but that makes sense. Maybe our diners will make unhealthy decisions about what to eat, which people outside the group will have to pay for in the form of medical bills. The wider society could have their own relatively minuscule say in the decision by voting in favor of a tax on unhealthy foods. A major weakness of this liberal democratic approach to meals is that it doesn't take into account the interests of creatures that can't participate in dining decisions. What about the animals our diners plan to eat?

tion as anyone to say how they'll affect you. In general, if we decide or contribute to decision making to our own benefit and we give people a proportional say in any decision, then we'll make decisions in a way that will work out best for the people who are affected by them most.

How could we apply this heuristic to a Martian colony? Well, to a first approximation we can say that the decisions concerning the use of some portion of land affect you more the closer you are to it. So give each colonist a certain number of ownership shares. Divide the surface of Mars up into a grid of lots. Take the proportion of the time in a year that an occupant of Mars spends in each lot. Half of their shares are divided among the lots they've visited in the year, proportional to the time they spend in each lot. The remaining shares are divided among the other lots, decreasing with distance, but in such a way that they still have at least one share in the most distant lots. Then everyone has ownership in a lot proportional to the extent that the decisions made by the owners affect them.

I mean, not really. This is a bonkers oversimplification of the actual problem. But it gets the conversation started. How would you improve on it?

> The girl who asked the Mars question came with her mom. While we talked with the mom for a few minutes, the daughter couldn't stop buzzing around—running, climbing, crouching, picking things up, putting them down. Eventually, the mom asked the daughter if she had any questions. They all came out at once: What would happen if we ever colonized Mars? Who would own the land? How could we settle disputes between different entities (individual workers? businesses? governments?) making legitimate claims to the land? What would make a price for the land fair or unfair? We didn't really know how to answer them, but she ran away before we could try.

How Do You Know That
$1 + 1 = 2$?

T his might seem like a bad question, because it might seem like the questioner is, in a distracting or annoying or lazy way, trying to call into doubt whether you actually do know that $1 + 1 = 2$. But even if we assume that this is something you know, there's still a real puzzle here. The problem is that you can't observe or perform experiments on numbers and other mathematical objects, at least not in the way that you can observe or perform experiments on electrons and stars and the medium-size dry goods that make up the objects of our everyday experiences. We seem to learn new mathematical truths often just by thinking. But how is that possible? After all, we can't learn much of anything else by thinking alone.

You might respond that we do actually know that $1 + 1 = 2$ from observation. Whenever we see one apple, then another apple, lo and behold, we see two apples, and so on. A more elaborate version of this point takes the form of what's called an indispensability argument. Our best scientific theories make reference to all sorts of mathematical objects and operations. We have very good evidence that these theories are true—or close to the truth—so we have good evidence that these mathematical objects are the way those theories say they are. The main problem with this proposal is that much of our mathematical knowledge doesn't have anything to do with physical quantities in the real world. We knew all sorts of things about complex numbers, for example, before we found any physical applications for them.

Another natural answer, if someone asked you how you know that $1 + 1 = 2$, would be to prove it from the axioms of arithmetic.* But this only kicks the problem down the road. How do you know the axioms of arithmetic?

In case it wasn't already obvious: I don't know. But one plausible answer is that the axioms of arithmetic are true by definition. You know them the same way you know that triangles have three sides, that bachelors are unmarried, and that vixens are foxes—by virtue of the meanings of the words. But there's a crucial difference between the axioms of arithmetic and these other definitional truths. It would still be true that all triangles have three sides, that bachelors are unmarried, that vixens are foxes, and so on, even if there were no triangles, bachelors, or vixens. The axioms of arithmetic, however, are true only if certain things actually exist—namely, the natural numbers. How could something be both true by definition and, so to speak, add to the list of things that exist?

Once again, I don't know, but I think it's worth exploring an analogy between numbers and things like corporations. If the right group of people get together under the right sorts of circumstances and declare that some new corporation exists, then it exists. We can bring corporations into existence just by talking about them in a certain way. But also, when we talk about corporations, we know that we're just talking about the activity of the people who make up the corporation. This is weird, but it's not completely mysterious. Maybe numbers and tensor fields and rings and all of the other things that

---

* The most commonly given axioms for arithmetic are the Peano axioms. You only need some of the Peano axioms to show that $1 + 1 = 2$. The Peano axioms use 0 as a constant and then define all of the natural numbers as successors of successors of 0: 1 is $S(0)$, 2 is $S(S(0))$, and so on. The two axioms of addition are that $x + 0 = x$ and that $x + S(y) = S(x + y)$. The proof that $1 + 1 = 2$, or that $S(0) + S(0) = S(S(0))$, is pretty simple. $S(0) + S(0) = S(S(0 + 0))$, by the second axiom of addition. And $0 + 0 = 0$, by the first axiom of addition. So $S(0) + S(0) = S(S(0))$.

mathematicians talk about work like that. (Although they can't be *exactly* like corporations, because corporations go into and out of existence and change over time, and numbers don't.) In any case, how we know about numbers might depend on what sorts of things numbers actually are.

Is Chicken Parmesan
Authentic?

Sometimes when I go on vacation, I just want to go to the beach or hike or sit around with my friends. But sometimes, especially when I'm going somewhere new, I want to get a sense of what life is like when the tourists aren't around. I want to do the things that locals do—hang out where they hang out, listen to the music they listen to, get a sense of the local politics, eat what they eat. Or rather, I want to do the *distinctive* things that they do; to the extent that the locals eat at Arby's and listen to Top 40 and watch superhero movies like everyone else, that's fine, but I'm not especially interested. (There's a tension here. Since globalization tends to smudge out cultural differences, I probably end up, more than I'd like, doing some sort of old-timey or ossified or caricatured version of the local thing. But I'm usually not too troubled by this.)

(I'm not sure *why* this is something people want to do on vacation. Maybe we're curious, maybe we like novelty, maybe we don't want to look like rubes, maybe it helps us imagine different ways our lives could have gone. I suspect it's a little bit of all of these.)

The thing is that when you're a tourist, people will try to sell you a version of local life that is made for tourists or adapted to their tastes. You'll go looking for the real thing only to find yourself in a crowd of other people looking for the real thing. So when you find the real thing, it's an accomplishment. This is one type of authenticity: what isn't made for outsiders, what's distinctive of the local way of life.

This type of authenticity is relative. Things aren't authentic full stop; they are authentic appurtenances *of* such-and-such culture. This is relevant to chicken parmesan, in particular. At some point, we find out that the Chinese/Italian/Mexican/whatever food that you get in most restaurants in the United States is not very much like the food that people eat in China/Italy/Mexico/wherever. For me, discovering this was disappointing. But should I have been disappointed? For starters, if the food is good, what does it matter if it's authentic? Not every experience is an attempt to familiarize yourself with another culture. The same point applies to all sorts of cultural products. There's a certain type of music listener who obsesses over whether something is *real* (that is, authentic) hip-hop or punk or blues. If your interest in music is purely ethnographic or historical, this obsession might make sense. But isn't this sort of interest weirdly narrow?

More to the point, though, even if it's not authentically Chinese/Italian/Mexican/whatever, couldn't it be authentically something else? If I wanted to experience the local food of Parma, I shouldn't be on the lookout for chicken parmesan. But if I wanted to experience the local food of northern New Jersey, chicken parm would be a great place to start. We could say the same thing about crab rangoon or hard-shell tacos. So on the one hand, worrying about whether something is or isn't authentic can be a distraction from other ways that it might be valuable. On the other hand, when you realize that authenticity is relative to a place or a way of life, you can find authenticity in places you might not have expected it.

But this isn't the only type of authenticity. You might spend time with someone and get the sense that they're putting on an act or that they're only behaving the way they are out of a sense that that's what's expected of them. They aren't being themselves (whatever that means). Just like you might want to know what some culture or way of life is really like, you might want to know how someone is when they aren't giving you what they think you want or otherwise worrying excessively about what you think about them. We use the language of authenticity to express these desires. So in addition to the cultural sort of authenticity at issue for the chicken parmesan consumer, we also evaluate individuals in terms of what we might call personal authenticity.

And just like demanding authenticity from some dish or piece of music might distract us from whether the dish or music is actually good, demanding authenticity from ourselves and other people can be exhausting and mean. Suppose that the cashier at the bodega isn't being himself, and he's just acting out the role of bodega cashier in his interactions with me. So what? Life is hard, his job is hard, and we both have more important things to worry about. Maybe in our most intimate relationships, where otherwise minor lapses of honesty can feel like deep betrayals, personal authenticity is more valuable. (Although even then: suppose I give my partner roses on Valentine's Day and she loves it. What does it matter if she's acting out the role of beloved girlfriend, and I'm acting out the role of the thoughtful boyfriend? Isn't it still nice?) In any case, if we should care about personal authenticity, we shouldn't care about it unconditionally, everywhere and always. If a little bit of mindless conformity or impression management or role-playing makes life easier for you and the people around you, cut yourself some slack.

Should You Kill Baby Hitler?

There are a lot of ways of approaching this question. We might treat it as a question about the great man theory of history. (If you killed baby Hitler, would the wider social forces that led to the rise of the Nazis have made up for it?) We might treat it as a question about how we evaluate historical counterfactuals in general. (How could you possibly know what would happen if you killed baby Hitler, since it's impossible to study these sorts of historical events experimentally?) We might even treat it as a question about the paradoxes of time travel. (If you killed baby Hitler, the delicate chain of events that led to your birth would be broken. But then how could you kill baby Hitler?)

I think most people have another interpretation in mind, though. Suppose that killing baby Hitler would have prevented the Holocaust (without causing anything comparably bad). Still, how could it be OK to kill a baby? Or, more generally, how could it be OK to punish someone for something they haven't even thought about doing yet? The short answer is simple: it would have been really great to prevent the Holocaust. The millions of lives you would have saved are more important than the life of baby Hitler. Killing baby Hitler would do an incalculable amount of good at relatively low cost.

✎ But isn't this just so much rationalization? It's simply wrong to kill babies and innocent people, and no amount of cost-benefit analysis is going to change that.

I agree that the feeling of revulsion at killing a baby doesn't go away even if you conclude that the benefits of killing baby Hitler outweigh the costs. I also don't want to just dismiss these sorts of powerful, automatic moral reactions. In our daily lives, we wouldn't get far without them. Of course, we shouldn't *always* trust these reactions. (Tons of people are disgusted by gay and interracial relationships; we shouldn't treat these feelings of disgust as knockdown arguments.) But if we're going to overrule our moral guts, we need some reason to do so. The cost-benefit analysis evidently isn't reason enough for some people.

Here's another way of thinking about it. Our basic repertoire of automatic emotional reactions evolved in a particular sort of ancestral setting and is molded in each of us by our cultural heritage and personal experiences. The reactions spread and persist, when they do, because they survived a process of trial and error. But we also carry these emotional reactions with us into situations in which they weren't tested. If we find ourselves in a situation that is different in some relevant way from the situations in which our moral instincts were tested, we shouldn't trust our moral instincts. It's as if something whose eyes evolved underwater were to come onto land and immediately start estimating the sizes and distances of things. It would do best to look past appearances.

Our revulsion at violence against babies evolved in a context where we needed to cooperate to raise children to maturity and in which babies required intensive investments of very scarce resources. Similarly, the special resentment we have toward people who hurt the innocent was necessary for us to be able to live side by side. (What's the point in being innocent—that is, cooperative—if you're just going to be punished anyway?) We react to violence against babies and innocent people in the way we do *because* our ancestors wouldn't have been able to survive and thrive if their neighbors didn't have those reactions themselves. But note how different the baby Hitler scenario is from our ancestral environment. Here, we know as a matter of stipulation that keeping the innocent baby alive will lead to millions of deaths. If we evolved in a world where we had to deal with baby Hitler scenarios all the time, our feelings about babies and the innocent would likely be different. The case is perfectly designed to take automatic moral reactions that normally sustain cooperative group life and make them, instead, incredibly deadly. If there's one case where our moral instincts shouldn't be trusted, it's this one.

But also: If you could kill baby Hitler, couldn't you prevent baby Hitler from being born in the first place? Or convince young Hitler to stick with his painting career? Or literally anything other than killing a baby? For the record, this idea came from another philosopher's comments[*] on a draft of this book. Maybe it says something about me and maybe it says something about doing philosophy through thought experiments, but it's a little troubling that it never occurred to me.

---

[*] Shout-out to Nancy McHugh.

Can Plants Think?

Plants can do some pretty wild things. They can move toward or away from certain stimuli, like light, water, heat, and touch. They can send signals through their root networks and through the air, preparing neighboring plants to defend against incoming threats. Different parts of the same plant can also communicate with one another when they're in distress. They can distinguish between relatives and strangers of the same species, cooperating with the former but competing for resources with the latter.

When people do these sorts of things, we often take it as evidence of beliefs and desires and other mental states. You recoiled because you didn't *want* to be touched. You reached for the top shelf because you *thought* you'd find cookies there. When you said "Look out!" you expressed your *desire* that your audience attend to a possible source of danger.

Then again, a lot of the most impressive plant behavior is pretty similar to what goes on in physical systems that we consider mindless. The mercury in an old-fashioned thermometer expands or contracts in response to changes in temperature. When something emits smoke, no matter how dumb it is, it signals to nearby creatures that it is on fire.

So are plants more like people or more like thermometers or burning pieces of paper? I'd say the latter, for a couple of reasons. First, while it's possible to use mind-y language to describe what plants do, it's also possible to use that same language loosely or metaphorically to describe mindless things. (The mercury "wants" to expand; the paper is "warning" us that it's on fire.) When is that language loose or metaphorical, and when is it strict or literal? One plausible answer is that it's literal when it allows us to predict and explain some behavior better or more succinctly than we could without it, with only tolerable losses of accuracy and precision. (Really, this is a continuum. The more we can predict and explain, better and more succinctly, the more literal; the greater the losses of accuracy and precision that come with the mentalistic language, the less literal.) Saying that a mercury thermometer wants to expand is pretty imprecise, it's not much of an explanation, and it's barely more succinct than saying that mercury expands at such-and-such rate under normal pressure. Mentalistic descriptions of plant behavior are roughly similar. It's not much of an explanation of (vocabulary word) phototropism to say that the plant wants to get closer to the light. We have more accurate, not particularly unwieldy explanations of this behavior in evolutionary and chemical terms.

Second, our thoughts about things are different from whatever plants do in at least two ways. One: our thoughts are *stimulus independent*. I can think about fried rice, say, both when it's in front of me and when I'm planning what to make for dinner tonight. A plant can send a signal in response to some threat, but can it think anything else about that threat when it's not there? Two: when you and I can think about something, we can think more than one thought about it. One way that you know I have the concept of an apple is that I not only believe a bunch of different things about apples (that they come in red and green varieties, that they grow on trees), but I can also have all sorts of other attitudes toward them (wanting to eat one, imagining a blue one). Philosophers sometimes call this the *generality constraint*. It seems to me that putative plant thoughts don't obey the generality constraint. A plant can move toward light, but can it think anything else about light unrelated to movement?

So I doubt plants can think, both because it's not clear what we get out of supposing that they can and because putative plant thoughts are suspiciously stimulus dependent and specific in content. We're tempted to attribute mental states to plants, I suspect, for the same reason that our ancestors were tempted to attribute mental states to winds and rivers and volcanoes—because our brains are working overtime to detect minds everywhere they can. In that way, seeing plant minds is sort of like seeing faces in our toast. We can't help but look for faces everywhere, and it's harmlessly exciting to find one in an unexpected place. But in our more sober moments, we realize that this is silly. Even if my breakfast looks exactly like the Virgin Mary, it's just an immaculate deception.

I'll show myself out.

Is Buddhism a Religion
or a Philosophy?

¿Porque no los dos? Think about some of the features that religions (tend to) have in common:

+ Distinctive ethical beliefs (keeping halal or kosher kitchens, prohibitions on taboo words)

+ Distinctive supernatural beliefs (God, the afterlife, reincarnation, Brahman)

+ The view that at least some of the central doctrines of the religion were revealed to a single sage (Moses at Sinai, Muhammad's night journey)

+ Holy texts or oral traditions (the Bible, the Egyptian *Book of the Dead*, the *Bhagavad Gita*)

+ Some (more or less disputed) methods for marking or defining membership in the religion (baptism, the doctrine of matrilineal descent in Judaism)

+ Some hierarchy of religious authority (monks, priests, imams, shamans)

+ Sacred rituals (prayer, meditation, life cycle ceremonies)

Buddhism comes in a lot of different forms, but most of them share all of these features. The Eightfold Path to liberation from the wheel of rebirth is, in large part, an ethical system; belief in re-incarnation is supernatural, and at least some Buddhists believe in a large pantheon of gods and otherworldly beings; Buddhists hold that much of their wisdom was revealed to the Buddha or at least that he discovered the central truths of Buddhism in an epiphany; different Buddhist sects have their holy texts, like the Pali canon; different Buddhist sects will count different people as their members, but the least controversial examples of Buddhists will be people who subscribe to the Four Noble Truths; there are Buddhist monks; there are Buddhist prayers, meditations, and other rituals. I'm sure you could come up with a passable definition of "religion" that excludes Buddhism and includes all the others, but it would have to be pretty gerrymandered. Why bother?

On the other hand, whenever I've spoken to Buddhists in the past, they've insisted that it's a philosophy not a religion.* Now, I'm not 100 percent sure what they have in mind. Maybe it's that they arrived at their Buddhist beliefs and practices as a result of careful reasoning or that they're prepared to defend them rationally or even that they're prepared to revise them in light of contradictory evidence. Maybe it's that certain Buddhist ideas—the empty self, the two truths, the ethical components of the Eightfold Path—are at least partial answers to philosophical questions. (What is the self? What makes me today the same person as me at age two? What is the fundamental structure of reality? How should I live?) Maybe it's that (some) people can adopt Buddhist beliefs and practices without leaving their own religions. Maybe it's just that calling a set of beliefs and practices a philosophy is, rightly or wrongly, a way of dignifying it. In any case, I don't disagree. Of course, young children raised as

---

* Some people have claimed that this whole trope is a response to Buddhism being exported to the Western world in some modernized form. Maybe, but it's interesting just how many people insist that Buddhism is a philosophy—both Buddhists of European descent raised in the United States and so-called ethnic Buddhists born and raised in Japan, in my own experience.

Buddhists didn't arrive at their beliefs through careful reasoning, and there are some versions of Buddhism with cosmological and theological tenets that are, on their face, inconsistent with the tenets of other religions. But if calling Buddhism a philosophy amounts to saying that the people who practice it reflectively and critically are thereby doing philosophy, then sure.

> ✎ But most Buddhists were raised as Buddhists in Buddhist communities. If someone holds a belief or acts in some way because that's what they were taught by their parents or because it's the thing that's done among their social group, it doesn't deserve to be called philosophical.

I agree that if someone just believes whatever their parents and their community tell them uncritically or unreflectively, calling that belief philosophical gives the wrong impression—like it's just as reasonable as anything that any philosopher could come up with.

But the "uncritically or unreflectively" is crucial, for a few reasons. First, philosophical reasoning tends to lead to the common sense of one's time and place. The ancient Greek philosophers mostly ended up agreeing with other ancient Greeks; medieval Christian philosophers were, you know, medieval Christians; for all of his elaborate reasoning, Kant's practical moral conclusions were garden-variety eighteenth-century Prussian Lutheranism.* And in each case, it's more or less impossible to say where the conventional wisdom ends and the philosophy begins. We still correctly consider all of these people philosophers.

Second, philosophical reasoning has to start somewhere. Why

---

* Friedrich Nietzsche, always good for a zinger: "Kant's Joke—Kant wanted to prove, in a way that would dumbfound the common man, that the common man was right: that was the secret joke of this soul. He wrote against the scholars in support of popular prejudice, but for scholars and not for the people."

should it matter whether the basic assumptions you're working from (at least provisionally) are things the people around you believe?

Third, we should distinguish having a philosophy from doing philosophy. You have a philosophy if you're prepared to answer some philosophical questions. To say that someone has a philosophy sounds like we're conferring a special status on some of their beliefs, but it shouldn't: you can have a philosophy even if it's incoherent or vague or ill considered. Doing philosophy, on the other hand, is a matter of reasoning about and struggling to refine and articulate your own and others' philosophical views. If a belief is the product of doing philosophy, it does have a sort of seriousness or respectability that not all beliefs share. Perhaps all Buddhists have a philosophy, but only some Buddhists come by their beliefs and practices by doing philosophy.

And finally, there's an unpleasant history of philosophers in the West ignoring anything written or thought by anyone outside of the European tradition on the grounds that it's *just* religion. If calling Buddhism a philosophy is what it takes to put that behind us, I'm down.

How Can Something Be
So Bad It's Good?

Some people like things in spite of or because of their badness: music like the Shaggs' album *Philosophy of the World* and Eurovision classics like "My Friend" or "Facebook Uh, Oh, Oh," movies like *The Room*, *Troll 2*, and *The Princess Switch*, and paintings like *Lucy in the Sky with Flowers*, and pretty much the whole collection of the Museum of Bad Art. Let's call these SBIGs.

But how is this possible? After all, we usually dislike art we think is bad. Isn't thinking something is bad just the same thing as disliking it?

At the booth where this question came up, the other philosopher said that a lot of SBIGs involve liking things ironically. That seems right to me, but it got me wondering what liking things ironically really is. After all, when I think about irony, usually I think about dramatic irony (where the audience has a much better understanding of what's happening to a character than the character themselves) and communicative irony (where, at least according to my preferred theory, someone ironically does something if they're trying to get you to see them as different from the sort of person who would do that thing in the circumstances). But the sort of irony involved in liking things ironically isn't, in general, dramatic or communicative; someone who likes something ironically can fully understand their own situation, and they can do it without communicating anything to anyone in particular. (More to the point, if you ironically thank someone for something, you don't actually thank them for it. But if you ironically like something, you actually like it.) Here's my working hypothesis: you like something ironically when you like it but doing so requires you to self-consciously and playfully suspend the standards you typically use to evaluate that sort of thing.

But I don't think that's the only way to get an SBIG. It's important that if SBIGs are good, they're good in a certain way. They're not, at least as far as I can tell, good in virtue of being elegant or profound or touching or insightful. Rather, they're good: (a) because they're funny[*] and (b) because they're flabbergasting. I feel an unusually intense desire to know how this thing managed to get made and to articulate just what makes it so weird.

This helps explain how learning more about an SBIG might complicate your feelings about it. After reading[†] that the Shaggs had pretty bleak childhoods and lives after the band, I felt uneasy about enjoying (?) their music, like I was punching down. But then I thought, conveniently: they themselves didn't enjoy their music very much, even if the liner notes claim the Shaggs love us and it wasn't their decision to put out their record. So I'm not laughing at them (at least not in any especially condescending way); I'm laughing at their music. There seems to be a morally important distinction between two types of SBIG—SBIGs whose makers realize that they've made something bad and SBIGs whose makers don't. In the latter case,

---

[*]   What makes SBIGs funny? Some philosophers have proposed that we find something funny when it causes us to regard ourselves as superior to other people (or past versions of ourselves). This would explain the funniness of SBIGs pretty well. Lucky for me, I don't have the right combination of grandiosity, creepiness, misogyny, and inattentiveness to the natural rhythms of human speech necessary to make *The Room*. (And if I did, I would be self-aware enough not to actually make the movie.) As a general account of what's funny, the superiority theory is wrong; if you take a minute, you can think of all sorts of counterexamples to it. But I think it's safe to say that sometimes we laugh because we regard ourselves as superior to someone. It's not clear, however, how the superiority theory would handle the fact that not all really bad art is SBIG.

[†]   Susan Orlean, "Meet the Shaggs," *New Yorker*, September 22, 1999, accessible at https://www.newyorker.com/magazine/1999/09/27/meet-the-shaggs.

part of what we're laughing at is the maker's lack of self-awareness, but in the former case, we might just be laughing at the work itself or laughing at features of the maker that aren't quite as touchy.

This is part of what makes me reluctant to describe kids' art as SBIG. Kids generally aren't especially self-aware about their own artistic skill, but it's mean to laugh at them for it.

Do Jelly Beans Taste the Same to People Who Like Them and to People Who Dislike Them?

There are some pretty persuasive arguments for both a yes and a no answer.

TEAM TASTES THE SAME	TEAM TASTES DIFFERENT
As one kid suggested at the booth: suppose you eat the same jelly bean at two different times. The first time, you enjoy it. The second time you don't, not because your tastes have changed or anything like that but because you're stuffed or you've eaten too much candy recently or you're just not in the mood. The jelly bean tastes the same, right? And if a jelly	If jelly beans taste the same to jelly bean lovers and jelly bean haters, we have to be able to distinguish the experience of enjoying how something tastes from the experience of tasting the thing itself. But it's not clear that that's possible. In addition to the experience of sweetness, mushiness, etc., there must be a separate experience of enjoying. What is that experience like? Is it the same thing you experience when you enjoy any food or enjoy anything at all? It seems odd to say that eating a delicious pickle

bean tastes the same to a single person across occasions when that person likes it and when they dislike it, why must it taste different to jelly bean lovers and jelly bean haters?

If you ask lovers and haters how jelly beans taste, I bet they'd describe them the same way—sweet, a little hard on the outside and mushy on the inside, a little grainy, they sort of dry your mouth out, etc. If that's true, the simplest explanation of why is that jelly beans taste the same to both groups of people.

is the same, in any meaningful way, as eating a delicious jelly bean: pickles and jelly beans are completely different.

Often, if one person likes some food and another person dislikes it, they'll describe it differently. People who dislike cilantro describe it as tasting soapy; people who like it don't. Clearly, cilantro doesn't taste the same to these people. Now, lovers and haters might not describe the taste of jelly beans any differently (other than by saying they like or dislike them), but that could just be because they lack the language to describe the difference in taste. So if, generally, one person likes some food and another dislikes it, it tastes different to them, then the same goes for jelly beans.

Where does this leave us? In a memorable section of his *Critique of Pure Reason*, the philosopher Immanuel Kant presents what at first appear (at least to him) to be fully convincing arguments for contradictory answers to some formidable metaphysical questions—for example, whether space is infinite and whether there is a "necessary being" (read: God) beyond the material world. He concludes that this is possible only because there's something wrong with the questions themselves.

The jelly bean question isn't quite as austere as the problems that exercised our boy Manny, but it might appear that we're basically in the same boat. For example, you might think the disagreement between the two teams is merely verbal. Everyone might agree about what it's like for jelly bean lovers and haters when they eat jelly beans; they just disagree about whether to describe the difference between those experiences as a difference in taste. If how something tastes includes whether you enjoy it, jelly beans taste different to lovers and haters. If it doesn't, they taste the same.

I don't think that's it, though. The question is whether there's always or necessarily a difference in what it's like for lovers and haters to eat jelly beans, *apart from* the difference in enjoyment itself. It's hard to say one way or the other, but it's not meaningless or hopelessly vague or merely verbal.*

In any case, the best I can give you is: I don't know, but I lean Team Tastes the Same. The first argument is just very compelling. It seems clear to me that when you eat too much of a good food, it goes from tasting good to tasting bad without necessarily changing taste. (It *might* change taste—say, if your mouth begins to burn from eating too much of something spicy or if it dries out from eating too much salt. But not necessarily.) While we never get the chance to be someone else and directly compare their experiences of eating jelly beans with our own, we can compare our own experiences with jelly beans over time, which is close enough to the same thing. The second argument is pretty weak, though. After all, it's notoriously hard for untrained people to describe tastes in general.

The arguments for Team Tastes Different don't hold up quite

---

* It's also worth noting that there are a lot of tricky questions about taste that I'm trying to dodge here. Conventional wisdom has it that the mouth is the seat of taste, but our experiences of taste are largely determined by what goes on in our noses. How much are our intuitions about taste affected by this mismatch between conventional wisdom and physiology? Smell, more than other senses, is closely related to emotionally laden memory. Does that mean that the relationship between taste and enjoyment is different than, say, the relationship between vision and enjoyment? Sometimes we describe the object of taste as a whole object (e.g., a jelly bean), sometimes as a quality of the object (e.g., tanginess). Is one of these the *real* object of taste, and if so, what are the consequences for the jelly bean question?

as well under scrutiny. The first argument rests on some intuitions about how we tease apart different elements or components of this sort of holistic experience—the experience of eating a jelly bean. These are intuitions about an abstruse psychological question, not something that we have regular experience with; they don't count for much. The second argument gets its appeal from the questionable assumption that all differences in food preferences are like the differences between people who like cilantro and people who dislike cilantro. But some food preferences are genetic (like the preference for cilantro), others we acquire by enculturation (like the preference for spicy food in certain regions), and others (like jelly bean preferences) are a matter of significant individual difference even within members of the same social groups. I don't see why we should expect the connection between taste and enjoyment to be the same for all of these types of food preference.

Isn't it interesting, though, that we have to reason this indirectly about our experiences? There's a venerable tradition in the history of philosophy according to which we only know about the external world on the basis of what we first learn about our own sensations. But the way these sorts of discussions go should, I think, call that tradition into question.

In conclusion, you're objectively wrong if you like jelly beans, except for the sour ones.

If You Try to Make a Sculpture of a Dragonfly and It Ends Up Looking Like a Bee, Is It a Sculpture of a Dragonfly or a Bee?

It's a dragonfly. The appeal of thinking that it's a sculpture of a bee derives, I think, entirely from the appeal of the resemblance theory of representation. According to the resemblance theory, one thing is a representation of another only if it resembles it. So if I imagine a rhinoceros, at least part of what makes it the case that I'm imagining a rhino—rather than a hippo—is that my mental image resembles a rhino. The resemblance theory has a distinguished list of supporters from the history of philosophy. But we know that it's wrong for at least two reasons.

First, lots of things represent things without resembling them. The color blue might represent states that vote Democrat on a map, but colors don't resemble any one political party more than any other. Or, to take my favorite example, have you ever had one of those dreams about, say, your childhood home, where the home in the dream didn't look anything like your actual childhood home? (In fact, the home might even look exactly like another building in the real world.) These dreams are interesting for all sorts of reasons I think, but one reason is that they undermine the resemblance theory.

Whatever makes it a dream *of* your childhood home can't be that the dream home resembles your childhood home.

Second, and more generally, a theory of representation needs to make room for the possibility of misrepresentation—that is, the possibility that one thing can be a representation of another, while attributing properties to the thing that it doesn't really have. The resemblance theory of representation seems to make it impossible for someone to misrepresent things too egregiously. Suppose that I take a commission to paint a portrait of George W. Bush. Not being a very skilled painter, I produce something that doesn't look anything like the former president but coincidentally bears a remarkable resemblance to a species of lemur I've never heard of. I haven't made a highly realistic painting of this lemur. I've made an inept painting of Bush. In other words, it's not that I haven't done the job at all but that I've done a bad job.

Now, this might convince you that the sculpture *isn't* a sculpture of a bee, but it doesn't tell us why it *is* a sculpture of a dragonfly. That's a trickier problem, but we can at least outline (what I take to be) the answer briefly. Consider the distinctive causal role that dragonflies play in the production of the sculpture. One day, we can suppose, someone saw a dragonfly and coined the word "dragonfly" to refer to creatures of that species. The word spread, and its meaning was refined by entomologists over the years. The usage of these entomologists eventually found its way to my doorstep. Now when I categorize things as dragonflies, either out loud or silently, I'm trying to do so in accordance with this usage. So there's a certain sort of causal chain that starts with actual dragonflies out there in the world, passes through language, and ends up in a sort of mental file that I have stored away in my brain, which I call up whenever I think of dragonflies. That mental file, in turn, plays a certain causal role in the production of the sculpture. If my sculpture resembles a bee on the other hand, that isn't because bees indirectly caused the sculpture to take shape. It's just because I'm not such a great sculptor. This yields an incredibly sketchy but decent explanation of the case at hand:

what makes it a sculpture of a dragonfly is the distinctive role that dragonflies played in the production of the sculpture. Good enough for government work, as my grandfather used to say.

But I'll note that this won't cut it, even in outline, as a general theory of representation. In particular, it will have a hard time dealing with representations of things that don't exist—mythical creatures, posits of antiquated theories, free lunches. Since these things don't exist, they have a hard time causing much of anything to happen. There's a shiny new PhD in philosophy waiting for whoever can figure out how we ever manage to represent nonexistent objects.

A mom and her son, who looked like he was about five, stopped by the booth. The mom talked to the other philosophers while I talked to the kid, who launched into an explanation of how to make a bumblebee out of pipe cleaners. The explanation went into some detail, so I had a minute to come up with the dragonfly question. When I put the question to him, all of a sudden he became a bit shy. His mom jumped in and asked the dragonfly question again. He got it right.

It's funny that people can feel perfectly comfortable talking with you about, say, pipe cleaners, but they get sort of squirmy when the conversation becomes philosophical. Was it just that the question was confusing? Was he afraid of saying something silly? Something else? Anyway, I guess it's cheating to include the question in the book, but I enjoy it. All the other questions came from the visitors, I swear.

What Makes Simpler
Theories Better?

I n our scientific theorizing about the world—but also in common sense—we prefer simpler theories to more complex ones. You can see this preference at work all over the place if you look for it. Back when we thought that the earth was the center of the universe and that all orbits were circular, astronomers had to add a bunch of geometrical bells and whistles to their models of the heavens in order to make sense of some seemingly contradictory observations.* One of the advantages of Copernicus's heliocentric theory was that it cut back considerably (though not entirely) on these additional gizmos. More recently, part of Noam Chomsky's influential critique of behaviorism in psychology was that behaviorists would have to posit an unwieldy number of "drives" in order to explain how animals learn. Or, to take a different sort of case, suppose a scientist gathers a bunch of data about two variables, plots them out, and sees that they take a rough U shape. But she also sees that there's a weird, squiggly curve, one that would take more math to describe, which fits the data exactly. If she's like most scientists, she'll say that the *real* relationship between the two variables is roughly U-shaped not the weirder shape.

This preference is perfectly natural but also more than a bit

---

* If you've ever heard someone say, pejoratively, that someone is adding an epicycle to some theory or debate, this is that.

mysterious. After all, the world itself isn't exactly simple. So if theories are supposed to describe and explain how some corner of the world works, it looks like a preference for simpler theories will often lead us to get things wrong.

So what, if anything, justifies this preference? One type of answer appeals to the fact that theories aren't *just* supposed to describe and explain how things work. For example, we prefer elegant or beautiful theories, and so we might prefer simpler theories because they're more elegant or beautiful. But that seems sort of frivolous. And if scientists cared that much about theoretical elegance and beauty, wouldn't they be better writers?

Less frivolously, but in the same vein, we might prefer simpler theories because they're easier to reason with. Theories are things we have to apply and test, and if a theory gets too complicated to apply or test, that makes it practically useless. This is true as far as it goes, but it doesn't go far enough. We are happy to accept theories that we don't yet know how to apply in a range of cases. (Hundreds of years after Newton, we still haven't solved the three-body problem: given the locations and velocities of three objects in a closed system, how, according to Newton's laws of motion and gravitation, will they move next?) And suppose that a theory says that the relationship between two variables is some hard-to-describe curve. It might not be practical for *me* to calculate the value of one variable, given the other, but that's what computers are for. If our preference for simpler theories were just a matter of ease of calculation, we would expect that preference to have weakened considerably since the advent of the calculator. But as far as I know, it hasn't.

In order to answer the question, we'll have to revisit the underlying premise that simplicity and the truth are at odds. There's at least one way in which simplicity and the truth go hand in hand. What philosophers call complex theories, statisticians and machine learning people call overfitting a model to the data. One problem with overfitting can be described in terms of the bias-variance tradeoff: roughly, the more accurately a model describes some set of data, the less accurately it will apply to new data. So one model might describe some sample perfectly using a million parameters, but it completely falls apart when we try to apply it to the wider population; while another model with a handful of parameters doesn't fit the sample so well, but it fits new data much better. This is one way of putting Chomsky's critique of behaviorism that I mentioned above: the more innate drives behaviorists have to posit in order to explain what they already know about learning, the less accurately they'll be able to predict what they don't already know about learning. (We can take the point further, too. Overfitting a model to some sample will get the model to include bad or noisy data, which we ultimately want to exclude. So a preference for simpler theories will also help us sort out good from bad data in our sample.) Our preference for simplicity prevents us from overfitting to what we already know, which will lead us to make more accurate predictions (and perhaps better understand the data we already have).

Again, though, this isn't the full story. Even if we have all the data we could ever want about some population, we might still prefer simpler theories to more complex ones. Since we already have all the data, the bias-variance tradeoff is beside the point. Maybe here we just fall back on the pragmatic, non-truth-related reasons to prefer simple theories from a couple of paragraphs ago? Or is there another connection between simplicity and truth that I've ignored so far?

If Superman Gets His Power from the Sun, Why Doesn't He Have a Tan?

Maybe there's no satisfactory answer to this question; the story just isn't fully consistent. There are some possible explanations of why he doesn't have a tan, though: he's a Kryptonian, and Kryptonians don't tan; getting a tan is a kind of skin damage, and he's impervious to most forms of skin damage; he does tan, but changes in his skin color are too subtle to represent in a comic or onscreen. Or maybe we could explain it by going outside of the story altogether: as one philosopher* has suggested to me, Superman is a product of a culture that values whiteness.

This is fun to think about. But I'm less interested in answering this guy's question than in why he posed it in the first place. IIRC, what happened is that he was watching a Superman movie, and he discovered this little incongruity, and the more he thought about it, the less invested he was in the movie. This raises a bona fide philosophical and psychological problem. Some things take us into stories. They lead us to respond emotionally to events in the story in roughly the way we would respond to those events in the real world. They lead us to identify with or see ourselves in certain characters, to simulate for ourselves what they're going through. Maybe they lead us to

---

* Shout-out to Nancy McHugh, again.

imagine what things look like or how they sound or feel. They lead us to care what happens next. They transport us, as psychologists say. But once we're in a story, we can be taken out of it. What we feel, what we imagine, what we want to find out, whose minds we read becomes less about what takes place in the story and more about what takes place outside of it. So the problem is: Why did the Superman tan question take this guy out of the story? Or, more generally, what draws people into a story and what takes us out?

(This is related to what philosophers sometimes call the paradox of fiction: given that people don't usually respond emotionally to what they know to be literally false and that fiction readers know that they're reading fiction, why do we respond emotionally to fiction? Like the problem I just raised, this is about explaining why people respond to stories in the way that they do and presumably a satisfactory solution to the paradox of fiction would tell us something about what, in general, draws us in or takes us out of a story. Still, our problem isn't quite the same as the paradox of fiction. First, the paradox is just about emotional responsiveness to fiction, not all of the other ways we're drawn into stories. Second, the paradox only asks us why we sometimes respond emotionally to fiction, not when we respond to fiction in general.)

You can think of a few different ways people can be taken out of a story: bad prose or acting, bad pacing, logical inconsistency, someone in the story acting out of character, "unearned" dramatic emphasis, boredom, difficulty following the plot. All of these things will affect different audience members differently; most people aren't taken out of the Superman movie by thinking about his tan. How exactly all of this works is a question for psychologists. But there is one way that we can be taken out of a story that philosophers have spent some time thinking about.

Here's a short story:

> Colonel Mustard was a nice old man, but in the afternoons he took to napping on a couch in the mansion's library. He snored terribly, great honking snores that could knock the books off

the shelves. Mrs. Peacock, who spent all her days reading in the library, couldn't concentrate at all. It annoyed her to no end; as soon as she got into a story, the colonel's snoring pulled her out of it. Then one day, she had a great idea. She mustered all of her bravery, crept up on the sleeping colonel, and hit him as hard as she could with a candlestick. His snoring had ceased at last! She tidied up to make sure that she wouldn't be caught, went back to her book, and soon disappeared into it. People were sad that Colonel Mustard had died, but Mrs. Peacock could finally read in peace. She knew in her heart that she had done the right thing, and she was right.

The story is written as if Mrs. Peacock were the hero. But there's something off about a story that stipulates this sort of thing. When you read the story, you might imagine Colonel Mustard snoring or Mrs. Peacock cleaning up the murder scene, but you don't imagine that it was *actually right* for Mrs. Peacock to kill her companion. Even though the story says that Mrs. Peacock was right to kill him, it seems like a stretch to say that it's *actually right* in the world of the story that she did the right thing. This is weird. We'll imagine and accept fictional worlds where all sorts of bizarre or improbable things happen—flying aliens with X-ray vision and super strength who shoot lasers out of their eyes—but if a story invites us to a world that's *morally* different from ours, we demur. Why is that? Philosophers call this little tangle the puzzle of imaginative resistance.* Or really, there are at least two puzzles, one psychological and one

---

* Some philosophers doubt that imaginative resistance ever happens in real stories, but I think it does. The Bible, for example, is chock full of stories of God doing horrible things to people; in 2 Samuel, the oxen carrying the Ark of the Covenant stumble, and when the guy walking next to the ark tries to steady it, God kills him. (Of course, because it's God doing this stuff, it's supposed to be part of the story that it's OK.) The Jewish tradition, at least, is to try to write a *midrash* or backstory that would justify God's actions. It's not a stretch to read the practice of midrash as, in large part, a response to imaginative resistance.

metaphysical: Why can't we so much as imagine that Mrs. Peacock did the right thing? And why is it not true in the world of the fiction that Mrs. Peacock did the right thing, even though that's what the story says?

One thing this shows is that our moral beliefs (and any other beliefs that give rise to imaginative resistance) are different in some important ways from our beliefs about ordinary matters of fact. Maybe it's that we guard our moral beliefs very carefully, so we're afraid to even imagine moral possibilities that might contaminate them. Maybe it's that we take the moral facts to be determined by the nonmoral facts, so that if a story tries to change the moral facts without making the necessary adjustments in the nonmoral facts, it will, in a manner of speaking, contradict itself. Or maybe (a more exciting possibility, in my opinion) moral beliefs are more like emotions or desires or plans than they are like garden-variety factual beliefs, and so imagining moral possibilities is just not the same sort of thing as imagining factual possibilities.

Is Anything Really Random?

We typically say that when you roll a fair six-sided die, it's random which side will turn up. This might mean a few different things, depending on the underlying interpretation of probability we're working with. On a strict frequentist interpretation of probability, what it means is that the die has been thrown a very large number of times, and each side has come up a roughly equal number of times. On a hypothetical frequentist interpretation, it means that *if* you threw the die a very large number of times, each side *would* come up a roughly equal number of times. On a subjectivist Bayesian interpretation, it means that you believe to degree 1/6 that it will come up on any one side on any given throw.* On an objectivist Bayesian interpretation of randomness, what it means, roughly, is that given the information you have at your disposal, an ideally reasonable person would believe to degree 1/6 that it will come up on any one side on any given throw.

A die can be random (*really* random!) in all of these senses. You might have some doubts about the counterfactuals involved in some of these interpretations, or you might doubt whether people's beliefs really come in degrees that can be assigned numerical values. But at

---

* "Bayesian" refers to the early probability theorist Thomas Bayes, the namesake of Bayes' theorem. Some Bayesians think of degrees of belief in terms of bets. If you believe to degree 1/6 that the next throw of the die will be a five, then you'll pay $1 for a bet that pays off $6 if it comes up five and nothing if it doesn't. But as a description of how actual people make bets, this is pretty dicey, pun intended.

least setting those doubts aside, all of these interpretations of randomness correctly describe actual dice out there in the world. The important thing to remember, which I wish somebody had told me when I was a very, very confused probability and statistics student, is that the mathematical theory of probability is an abstract structure, which we can use to describe in more or less idealized ways all sorts of things in the real world.

But I have the feeling that this isn't going to satisfy the person who asked the question. Take the strict frequentist interpretation of randomness above. It's true that if I throw the die a bunch of times, each side will come up roughly equally often. (At least for most fair dice.) But suppose I throw the die and it comes up five. Then if I were to throw the die with *exactly* the same force, so that it turned and bounced exactly the same way, wouldn't it always come up five? If so, then the throw of the die is, in a sense, not really random.

So is anything really random, in that sense? In other words, is there any situation that if we repeated it exactly* a bunch of times some variable in the situation would turn out differently each time?

I think the answer is yes. One popular example of a process that is really random in this sense is radioactive decay. If you take a bunch of identical samples of uranium-238 and wait for them to decay, the atoms will decay in a different order each time. The order is completely unpredictable. This is as random as it gets. Adherents to the hidden variable theory of quantum mechanics will object that there has to be *some* way that each of these samples differ; there has to be some "hidden variable" that determines which atom is going to decay next. But this is, as far as I can tell, just an article of faith.

I've described the deep randomness involved in radioactive decay in frequentist terms. But frequentism only makes sense of random-

---

* What does it mean to repeat a situation exactly? Describe everything about a situation that affects how it will unfold. Any other situation that also meets that description repeats it exactly. More or less.

ness in repeated events. Some random events—say, that this particular atom will decay in the next five seconds—are one-off affairs. If not frequentism, what interpretation of probability best describes that randomness?

What Makes a
Work Derivative?

Every new piece of art is derivative to some extent; even the most original work copies or borrows or is shaped by the maker's knowledge of its predecessors somehow or other. The question is what makes something *too* derivative—derivative enough to be a problem.

Well, what sort of problem? The aesthetic problem, at least, is that derivativeness can devalue a work. This could be a very literal, financial sort of thing: people will pay much more for the original painting than for the reproduction. But more interestingly, it can also be a matter of whether we enjoy something or how we judge it. When Nickelback released their song "Someday," an enterprising music student named Mikey Smith noticed that it sounded a lot like their previous hit, "How You Remind Me." To prove just how similar the two songs are, he combined them into a mash-up, which after doing the rounds on the internet came to be called "How You Remind Me of Someday." Different people react to the mash-up differently, but at least for some listeners, the mash-up turned them off—to one or both songs or to Nickelback as a whole.

It seems to me that all this has something to do with our perception of the aesthetic virtues and vices (the aesthetically good or bad traits) of the artists themselves. If a band copies one of its own previous hits, we're liable to see them as one-trick ponies, unadventurous, lazy, or hacky. Alternatively, if an artist copies some other artist's work (not when making a reproduction of something else, but when *trying to make an original work*), that might be a reflection of the artist's excessive susceptibility to fads or trends. (Then again, if an artist samples some recording, in a way that reflects the breadth of their musical knowledge or their knack for making old ideas work in new contexts, that's not derivative.) Our appreciation of a piece of art is largely (though not entirely) a matter of what it tells us about the virtues and vices of the artist themselves, so this matters. This is one way, then, that a work can be too derivative—if the derivativeness is a reflection of an aesthetic vice of the artist.

An artist who visited the booth changed my mind about this stuff. I think I said something about how the whole system of copyrights and patents struck me as completely broken, and she told me that patents did something a little unusual for her. Every time she started a new series or devised a new technique, she tried to patent it. And it worked! She actually held a bunch of patents for her art. But while she did sell her stuff, she didn't seek out these patents for financial reasons. (She never actually enforced her patents.) What she wanted was some kind of official credit for the originality of her work, which she didn't think the art world (or copyright, apparently) was in a position to provide. I wonder how many artists and inventors are like her—people who just want to make a living doing what they do, who care about their "intellectual property" only to the extent that it gets them recognition for their work.

What's Your Favorite Animal?

There's a fun game you can play with Wikipedia. Go to a random page. Click the first link in the page to another Wikipedia article. Repeat. Count how many clicks it takes to get to the page for Philosophy.

(I just played it. My journey went: Neodactylota, Gelechiidae, Moths, Insect, Latin, Classical Language, Language, Linguistic System, Ferdinand de Saussure, Switzerland, Sovereign State, International Law, Nation, Community, Level of Analysis, Social Science, Knowledge, Discipline (academia), Fact, Reality, Object of the Mind, Object (philosophy), Philosophy. My score was 22.)

Part of what the Wikipedia game shows is that every subject has some connection to philosophy, if you're willing to look hard enough. The Wikipedia game is slightly misleading, though. The way it usually works is that each article is typically somehow more abstract than the last, and once you get abstract enough, you're talking about philosophy. But this isn't the only way to make a connection to philosophy. Philosophy is just as likely to come out of our encounters with the weirder parts of the concrete world.

That's where the favorite animal question comes in. One way to answer it is to try to think of the most philosophically provocative animal, the animal that points the way to the most interesting philosophical question or conclusion. When I was asked this question at the booth, my first answer was: butterflies. Caterpillars can learn things about their environments (e.g., about sources of food), go into a cocoon, emerge in butterfly form, and still remember what they learned when they were caterpillars! Not only that, but when caterpillars are in the cocoon, they don't just transform piece by piece into butterflies—they turn into goo, which in turn grows into a butterfly. But this means that a single memory or piece of knowledge can take three completely different physical forms—caterpillar, goo, butterfly. I don't think this means that caterpillars have souls or anything like that, but it does mean that a single mental state can be implemented or realized by three completely different physical systems. It is beautiful and bizarre and incredible that this is how the world works.

I could also have gone with a biologically immortal species of jellyfish. Would I like to live like that? What would I gain and what would I lose? Is anything like a human life compatible with this sort of biological immortality?

I could also have gone with coral reefs. A coral reef is a single organism. How is that possible? We have all sorts of symbiotic and dependent relationships with the organic and inorganic parts of our environments, just like corals. Countless organisms that don't share our DNA live in our guts and on our faces. Do human beings stop and start in space and time where we ordinarily think they do, or are we more like coral reefs, expanding indefinitely? What could settle the matter one way or another?

But those are just my examples. Exercise for the reader: come up with your own.

> Before he asked about my favorite animal, this kid asked whether I had ever played the video game *Road Rage*. I guess I could have asked him what he likes about the video game and turned it into a discussion about aesthetic judgment or whether video games are art, but I wasn't quick enough. Maybe it's always possible, but it's not always easy to draw a conversation back to philosophy.

Bonus Question

What's the Best Way to Teach Myself Philosophy? How Do I Start?

Read this book.

If, for some reason, that's not enough, check out the recommendations in the next section for further reading/viewing/listening.

If *that's* not enough, there are all sorts of ways to dive deeper. There are tons of good philosophy podcasts, like *Hi-Phi Nation*, *Examining Ethics*, *Elucidations*, *The History of Philosophy without Any Gaps*, and *New Books in Philosophy*. On YouTube, you can check out the *Wi Phi* channel or search for Bryan Magee's interviews. There are also some great feature-length movies about philosophy, like Astra Taylor's *What Is Democracy?* and Raoul Peck's *The Young Karl Marx*.

"Many of us learn best with other people. Brooklyn Public Philosophers, the public philosophy event series I organize, has a speaker series and a Facebook page and a bunch of other things going on, in addition to the Ask a Philosopher booth. If, like most people, you're not in New York City, you can check out the website of any nearby college's philosophy department for talks and other events. And if there's no philosophical community in your area, you can start one! Consider reaching out to SOPHIA (the Society of Philosophers in America) to get help setting up a philosophy discussion group."

Some people think that the best way to get into philosophy is to read some historical philosopher's most famous tome—Plato's *Republic*, Hume's *A Treatise of Human Nature*, Kant's *Critique of Pure Reason*, etc. My friend in high school used to call these the fatties. But the fatties are often harder to understand than they need to be, and these days philosophers communicate with each other primarily through essays. I'd recommend picking up a good introductory essay anthology, like the *Norton Introduction to Philosophy*.

People sometimes ask me what my favorite philosophy books are. I'm not sure, but the short list would contain: J. L. Austin, *Philosophical Papers I*; Rudolf Carnap, *The Logical Structure of the World*; Paul Grice, *Studies in the Way of Words*; Carl Hempel, *Aspects of Scientific Explanation*; Saul Kripke, *Naming and Necessity*; W. V. O. Quine, *Ontological Relativity and Other Essays*; and Mark Wilson, *Wandering Significance*.

Embarrassingly, that list is made up entirely of white guys, so let me add that some of my favorite pieces of philosophy are by women: Louise Antony, "The Socialization of Epistemology," in *The Oxford Handbook of Contextual Political Analysis*; Nancy Bauer, "Pornutopia," in her *How to Do Things with Pornography*; Elizabeth Camp, "Why Metaphors Make Good Insults," *Philosophical Studies* 174, no. 1 (2017), 47–64; Angela Davis, *Are Prisons Obsolete?*, chaps. 3 and 6; Ruth Millikan, "Pushmi-Pullyu Representations," in her *Language: A Biological Model*; Martha Nussbaum, "Love's Knowledge," in her *Love's Knowledge*, and "Objectification," *Philosophy and Public Affairs* 24, no. 4 (1995), 249–91; and Rivka Weinberg, "Why Life Is Absurd," *The Stone*, accessible at https://opinionator.blogs.nytimes.com/2015/01/11/why-life-is-absurd/.

There is, of course, also plenty of great philosophy from the non-Western world. You might check out: Avicenna's "floating man" argument, accessible at https://www.davidsanson.com/texts/avicenna-floating-man.html; Alexus McLeod, *Philosophy of the Ancient Maya: Lords of Time*; the selections from Mengzi, Mozi, and Zhuangzi in

Philip Ivanhoe and Bryan Van Norden's *Readings in Classical Chinese Philosophy*; Nagarjuna, *The Fundamental Wisdom of the Middle Way*, trans. Jay Garfield, chap. 24; Sebastian Purcell, "What the Aztecs can teach us about happiness and the good life," *Aeon*, accessible at https://aeon.co/ideas/what-the-aztecs-can-teach-us-about-happiness-and-the-good-life; Sogoyewapha, "The Religion of the White Man and the Red," accessible at https://www.bartleby.com/268/8/3.html; *The Classic of Filial Piety* on the relationship between the family and the state, accessible at http://chinesenotes.com/xiaojing/xiaojing001.html; *The Questions of King Milinda* on the self, excerpts accessible at https://www.budsas.org/ebud/ebsut045.htm; and Zera Yacob's *Hatata*, excerpts accessible at http://www.alexguerrero.org/storage/Zera_Yacob.pdf.

References and
Recommendations

For an exploration of some possible answers to the question why there is anything at all, which are all wrong: Jim Holt, *Why Does the World Exist?*

For the psychology of meaningful work: Barry Schwartz, *Why We Work.*

For an introduction to the effective altruism movement and a guide to the most important jobs there are to do in the world: visit http://80000hours.org.

For the connection between alien farmers and the meaning of life: Thomas Nagel, "The Absurd," *Mortal Questions*, accessible at https://philosophy.as.uky.edu/sites/default/files/The%20Absurd%20-%20Thomas%20Nagel.pdf.

For a related argument that life is absurd, which is also my single favorite piece of philosophy written for a general audience: Rivka Weinberg, "Why Life Is Absurd," *New York Times*, January 12, 2015, accessible at https://opinionator.blogs.nytimes.com/2015/01/11/why-life-is-absurd/.

For the idea that we don't know very much about our own experiences: Eric Schwitzgebel, "The Unreliability of Naive Introspection," *Philosophical Review* 117 (2008): 245–73, accessible at https://faculty.ucr.edu/~eschwitz/SchwitzPapers/Naive1.pdf.

For the idea that our knowledge of the external world is based on abduction: Bertrand Russell, *The Problems of Philosophy*, chap. 2, accessible at https://www.wmcarey.edu/crockett/russell/ii.htmError! Hyperlink reference not valid. https://www.gutenberg.org/files/5827/5827-h/5827 -h.htm.

For an argument that it is good to have children who will have good lives overall and for the philosophy of parenting more generally: Jean Kazez, *The Philosophical Parent*.

For a detailed exposition of the genealogical argument against managerial capitalism: Elizabeth Anderson, *Private Government*.

For how our thinking about color came to have a subjective and an objective dimension and why this is philosophically important: Zed Adams and Matt Teichman, "Episode 95: Zed Adams discusses the genealogy of color," *Elucidations*, April 10, 2017, accessible at https://lucian .uchicago.edu/blogs/elucidations/2017/04/10/episode-95-zed-adams -discusses-the-genealogy-of-color/.

For a defense of the possibility of time travel, which gets into some other fun questions in the philosophy of time: Ted Sider, "Time," in Earl Conee and Ted Sider, *Riddles of Existence*.

For Mengzi's fuller and rather more optimistic account of human nature: Matthew Walker, "Ancient: Mengzi (Mencius) on Human Nature," *Khan Academy*, December 26, 2014, accessible at https://www.khanacademy .org/partner-content/wi-phi/wiphi-history/wiphi-ancient/v/history-of -philosophy-mengzi-on-human-nature.

For an appropriately skeptical account of the idea of the true self or how we are deep down: Nina Strohminger, Joshua Knobe, and George Newman, "The True Self: A psychological concept distinct from the self," *Perspectives on Psychological Science* 12 (2017): 551–60, accessible at http://ninastrohminger.com/papers.

For a discussion that ties together our practical concerns and the concerns of scientists who study well-being: Anna Alexandrova, *A Philosophy for the Science of Well-Being*.

For an account of the historical relationship between science and religion, which is the basis for what I say here: John Hedley Brooke, *Science and Religion: Some Historical Perspectives*.

For a defense of an emotional state theory of happiness, which really shaped my thinking about this stuff: Dan Haybron, "The Nature and Significance of Happiness," in Susan Boniwell, Ilona David, and Amanda Conley Ayers, eds., *The Oxford Handbook of Happiness*.

For an empirically and theoretically richer discussion of the problem of absolute space: Nick Huggett and Carl Hoefer, "Absolute and Relational Theories of Space and Motion," *The Stanford Encyclopedia of Philosophy*, accessible at https://plato.stanford.edu/entries/spacetime-theories/.

For a gentle introduction to how explanations work and the psychological effects of explaining things: Tania Lombrozo, "The structure and function of explanations," *Trends in Cognitive Sciences* 10, no. 10 (2006): 464–70, accessible at https://cognition.princeton.edu/sites/default/files/cognition/files/tics_explanation.pdf.

For Aristotle's definitely real and not a joke lost manuscript on the philosophy of trolling: Rachel Barney, "[Aristotle], *On Trolling*," *Journal of the American Philosophical Association* 2, no. 2 (2016): 1–3, accessible at https://philpapers.org/archive/BARAOT-9.

For the idea that emotions represent whatever they were "set up to be set off" by: Jesse Prinz, *Gut Reactions: A Perceptual Theory of Emotion*.

For how we know when we're in love: Martha Nussbaum, "Love's Knowledge," in her *Love's Knowledge: Essays on Philosophy and Literature*.

For a general introduction to the philosophical questions raised by homosexuality: Brent Pickett, "Homosexuality," *The Stanford Encyclopedia of Philosophy*, accessible at https://plato.stanford.edu/entries/homosexuality/.

For an account of the metaphysics of gender: B. R. George and R. A. Briggs, "Science Fiction Double Feature: Trans Liberation on Twin Earth" (forthcoming).

For a deeper discussion of the accounting metaphor and other metaphors that guide our moral thinking and discourse: George Lakoff, *Moral Politics*.

For an account of what gentrification is, what harms it involves, and what nearby social problems we should pay more attention to: Ronald Sundstrom, *Gentrification, Integration, and Racial Equality* (forthcoming).

For a couple of the classic arguments that your death isn't bad for you: Epicurus, "Letter to Menoeceus," in *The Essential Epicurus*, trans. Eugene O'Connor.

For a few more of those arguments: Book 3 of Lucretius, *On the Nature of Things*, trans. Martin Ferguson Smith.

For the approach to the problem of personal identity (what makes you today the same person as you at age two?) that has helped me deal with my own death anxiety: Derek Parfit, "Personal Identity," *Philosophical Review* 80, no. 1 (1971): 3–27, accessible at http://home.sandiego.edu /~baber/metaphysics/readings/Parfit.PersonalIdentity.pdf.

For a more rigorous but also way more interesting critique of the idea that we have robust moral obligations to dead people, which also happens to be the first episode of my favorite philosophy podcast: S01E01, "The Wishes of the Dead," *Hi-Phi Nation*, accessible at https://hiphination.org /complete-season-one-episodes/episode-one-the-wishes-of-the-dead/.

For the essay from which I borrowed the idea that retirement and aging can teach us about vulnerability and the limits of our habitual ways of thinking: Jan Baars, "Aging: Learning to Live a Finite Life," *Gerontologist* 57, no. 5 (2017): 969–76.

For an elaboration and defense of the theory that mental illness is a harmful mental dysfunction: Jerome Wakefield, "The Concept of Mental Disorder," *American Psychologist* 47, no. 3 (1992): 373–88.

For a good explanation of the idea of the veil of ignorance and what it has to do with the idea of a fair contract: Michael Sandel, *Justice*, chap. 6.

For a slightly more technical discussion of trust in experts, which gets into a lot of fun philosophical problems that don't arise at this book's level of breeziness: David Coady, *What to Believe Now*, chap. 2.

For a deeper discussion of some aims of moral education and a bit of evidence supporting the suggestions I offer here: Bart Engelen, Alan Thomas, Alfred Archer, and Niels van de Ven, "Exemplars and nudges: Combining two strategies for moral education," *Journal of Moral Education* (2018): 1–20, accessible at https://doi.org/10.1080/03057240.2017.1396966.

For a more comprehensive discussion of the forms that unjust discrimination or oppression can take: Iris Marion Young, "The Five Faces of Oppression," in her *Justice and the Politics of Difference*, accessible at https://www.sunypress.edu/pdf/62970.pdf.

For a more pessimistic take than mine about the moral and institutional problem of climate change—which, if it's wrong, it's really important to figure out where and why it's wrong: Stephen M. Gardiner, "A Perfect Moral Storm: Climate Change, Intergenerational Ethics and the Problem of Moral Corruption," *Environmental Values* 15 (2006): 397–413, accessible at http://ww.hettingern.people.cofc.edu/Environmental_Philosophy_Sp_09/Gardner_Perfect_Moral_Storm.pdf.

For an argument that all temporal discounting is irrational: Preston Greene and Meghan Sullivan, "Against Time Biases," *Ethics* 125, no. 4 (2015): 947–70.

For the use of "weight" in response to the ketchup question and a general account of how word meanings that seem stable vary in response to changes in how we measure things: Mark Wilson, *Wandering Significance*.

For a good introduction to the philosophical problems raised by property: Jeremy Waldron, "Property and Ownership," *The Stanford Encyclopedia of Philosophy*, accessible at https://plato.stanford.edu/entries/property/.

For a fun, mind-blowing problem in the philosophy of math: Eugene Wigner, "The Unreasonable Effectiveness of Mathematics in the Natural Sciences," *Communications in Pure and Applied Mathematics* 13, no. 1 (1960), accessible at https://www.dartmouth.edu/~matc/MathDrama/reading/Wigner.html.

For a wild, vertiginous take on some different ways of thinking about authenticity and how they can come into conflict: Rebecca Roanhorse, "Welcome to Your Authentic Indian Experience™," *Apex*, August 8, 2017, accessible at https://www.apex-magazine.com/welcome-to-your-authentic-indian-experience/.

For the nature and significance of the distinction between automatic and deliberate moral judgment: Joshua Greene, "Beyond Point-and-Shoot Morality," *Ethics* 124, no. 4 (2014): 695–726, accessible at https://psychology.fas.harvard.edu/files/psych/files/beyond-point-and-shoot-morality.pdf.

For a creepy, suggestive movie about our revulsion at hurting the innocent: Narciso Ibáñez Serrador's *Who Can Kill a Child?*

For an eye-opening defense of the idea that whether something has beliefs is a matter of how succinctly we can describe it in belief-y terms: Daniel Dennett, "Real Patterns," *Journal of Philosophy* 88, no. 1 (1991): 27–51, accessible at https://ruccs.rutgers.edu/images/personal -zenon-pylyshyn/class-info/FP2012/FP2012_readings/Dennett _RealPatterns.pdf.

For a defense of a similar conclusion about Buddhism from someone who knows a lot more about Buddhism than I do: Evan Thompson, *Why I Am Not a Buddhist.*

For an attempt to make some trouble for a commonsense understanding of consciousness, using a variation on the jelly bean example and some related evidence: Daniel Dennett, "Quining Qualia," in Marcel and Bisiach, eds., *Consciousness in Modern Science*, accessible at http://cogprints .org/254/1/quinqal.htmError! Hyperlink reference not valid.

For an introduction to the philosophy and science of smell (and a bit about taste): Ann-Sophie Barwich, "Making Sense of Scents: The Science of Smell," *Auxiliary Hypotheses*, accessible at https://thebjps .typepad.com/my-blog/2017/01/making-sense-of-scents-the-science -of-smell-ann-sophie-barwich.html.

For a fascinating discussion of alief, a psychological state kind of like belief, which might have a role to play in solving the puzzle of imaginative resistance, the paradox of fiction, and the problem of transportation in general: Tamar Gendler, "Alief and Belief," *Journal of Philosophy* 105, no. 10 (2008): 634–63.

For a relatively easy introduction to the different interpretations of probability and their strengths and weaknesses: Hugh Mellor, *Probability: A Philosophical Introduction.*

For the classic discussion of the significance of artistic reproduction: Walter Benjamin, "The Work of Art in the Age of Mechanical Reproduction," reprinted in his *Illuminations*, trans. Harry Zohn, accessible at https:// www.marxists.org/reference/subject/philosophy/works/ge/benjamin .htm.

For a story that manages to both dramatize copyright law and make serious philosophical progress on the topics of derivativeness and originality: Spider Robinson, "Melancholy Elephants," *Analog*, June 1982, accessible at http://www.spiderrobinson.com/melancholyelephants.html.

For the philosophical significance of animals that have weird boundaries in space and time: Derek Skillings, "Life is not easily bounded," *Aeon*, accessible at https://aeon.co/essays/what-constitutes-an-individual-organism-in-biology.

Acknowledgments

My name is on the cover, but this book is a collective effort. The Brooklyn Public Library has helped fund and schedule the Ask a Philosopher booth and has otherwise been a tremendous friend to philosophy in New York City. We've also received funding from the American Philosophical Association's Berry Fund and Humanities NY. The booth has been hosted by GrowNYC's Greenmarkets, Turnstyle Underground Market, the Brooklyn Book Festival, Socrates Sculpture Park, Bryant Park, Brooklyn Pride Parade, IMPACCT Brooklyn, the Flatbush Avenue Street Fair, West Elm, City Point, the Market at the Brooklyn Museum, SEPTA, the Metropolitan Transportation Authority, and (in radio form) by WNYC's *All of It* with Alison Stewart. The premise of the book came from my editor, Stephen S. Power, without whom not. Endless gratitude to all the philosophers who have contributed their time, energy, and expertise to the Ask a Philosopher booth: Leslie Aarons Stewart, Ben Abelson, Ericka Abraham, Zed Adams, Roman Altshuler, Carlo Alvaro, Vinny Andreassi, Elvira Basevich, Margaret Betz, Joe Biehl, Carrie Ann Biondi, D Black, Adam Blazej, Michael Brent, Evan Butts, Cristina Cammarano, Kristi-Lynn Cassaro, Kevin Cedeño-Pacheco, Ignacio Choi, Skye Cleary, Jesi Taylor Cruz, Zoe Cunliffe, Henry Curtis, Ryan Felder, Phoebe Friesen, Kate Ghotbi, Anna Gotlib, Dana Grabelsky, Pamela Guardia, Alex Guerrero, Bixin Guo, Noah Hahn, Ethan Hallerman, Geoff Holtzman, Brian Irwin, Marilynn Johnson, Jenny Judge, Justin Kalef, Laura Kane, Jonathan Kwan, Arden Koehler, Zoey Lavallee, Céline Leboeuf, Qrescent Mali Mason, Andrew McFarland, Lee McIntyre, Joshua Norton, Claudia Pace, Connie Perry, Jeanne Proust, Qianyi Qin, Shivani Radhakrishnan, Rick Repetti, Bryan Sacks, Greg

Salmieri, Miriam Schoenfield, Damion Scott, Jennifer Scuro, Casandra Silva Sibilin, Bart Slaninka, Joanna Smolenski, Alex Steers-McCrum, Christopher Steinsvold, Ali Syed, Travis Timmerman, Quixote Vassilakis, Denise Vigani, Paul de Vries, Thomas Whitney, Matthew Young, and anyone else I may have missed. Nancy McHugh and Derek Skillings offered helpful comments on drafts of the book. My partner, Jen Ortiz, and my parents, David Olasov and Sharon Spellman, have kept the booth running in ways both material and spiritual. And of course, this book is animated by the ideas, concerns, and quirks of every person who has stopped by the booth to chat. It is for you.